Dr. med. Dipl. Biol. M. Pruggmayer
Fachärzt f. Humangenetik
Facharzt f. Frauenheilkunde
Bahnhofstr. 5 • D-31224 Peine
T. 05171 / 37 75 • Fax 06171 / 1 21 71

HUMAN CYTOGENETICS

ESSENTIAL DATA SERIES

Series Editors

D. Rickwood
Department of Biology, University of Essex,
Wivenhoe Park, Colchester, UK

B.D. Hames
Department of Biochemistry and Molecular Biology,
University of Leeds, Leeds, UK

Published titles

Centrifugation
Gel Electrophoresis
Light Microscopy
Vectors
Human Cytogenetics

Forthcoming titles

Animal Cells: culture and media
Nucleic Acid Hybridization
PCR
Enzymes in Molecular Biology
Transcription Factors
Protein Purification
Immunoassays

See final pages for full list of titles and order form

HUMAN CYTOGENETICS
ESSENTIAL DATA

Edited by

D.E. Rooney
Cytogenetic Services, London, UK

B.H. Czepulkowski
Cytogenetic Unit, Department of Haematology, King's College Hospital, London, UK

JOHN WILEY & SONS
Chichester · New York · Brisbane · Toronto · Singapore

Published in association with BIOS Scientific Publishers Limited

©BIOS Scientific Publishers Limited, 1994. Published by John Wiley & Sons Ltd, Baffins Lane, Chichester, West Sussex PO19 1UD, UK, in association with BIOS Scientific Publishers Ltd, St Thomas House, Becket Street, Oxford OX1 1SJ, UK.

All rights reserved. No part of this book may be reproduced by any means, or transmitted, or translated into a machine language without the written permission of the publisher.

British Library Cataloguing in Publication Data
A catalogue record for this book is available from the British Library.

ISBN 0 471 95076 9

Typeset by Marksbury Typesetting Ltd, Bath, UK
Printed and bound in UK by H. Charlesworth & Co. Ltd, Huddersfield, UK

The information contained within this book was obtained by BIOS Scientific Publishers Limited from sources believed to be reliable. However, while every effort has been made to ensure its accuracy, no responsibility for loss or injury occasioned to any person acting or refraining from action as a result of the information contained herein can be accepted by the publishers, authors or editors.

Library of Congress Cataloging in Publication Data
Human cytogenetics: essential data/ edited by D.E. Rooney, B. Czepulkowski.
    p. cm.—(Essential data series)
    'Published in association with BIOS Scientific Publishers Limited.'
    Includes bibliographical references and index.
    ISBN 0-471-95076-9: £12.95
    1. Human cytogenetics—Tables. 2. Medical genetics—Tables.
  I. Rooney, D.E. (Denise E.). II. Czepulkowski, B.H. (Barbara H.) III. Series
    [DNLM: 1. Chromosome Abnormalities—diagnosis. 2. Karyotyping. 3. Cytological Techniques. 4. Genetics, Medical. QS 677 H918 1994]
  QH431.H8345 1994
  573.2'1——dc20
  DNLM/DLC
  for Library of Congress
                            94-30736
                                CIP

CONTENTS

Contributors	x
Abbreviations	xii
Preface	xv

1. Constitutional chromosome abnormalities.
J. Wolstenholme and I.E. Cross — **1**

Figures and Tables

The human karyotype	2
Incidence of constitutional abnormalities in spontaneous losses and CVS	5
Incidence of constitutional abnormalities at amniocentesis and birth	7
Karyotype variation in Turner syndrome	9
Parental origin of common constitutional chromosome abnormalities	10
Incidence of structural chromosome abnormalities	11
Incidence of rarer structural chromosome abnormalities	11
Robertsonian translocations – trivalent segregation, e.g. t(14q21q)	12
Segregation patterns of Robertsonian translocations at amniocentesis and CVS	13
Reciprocal translocations – quadrivalent segregation	14
Reciprocal translocations – quadrivalent segregation (figure)	15
Meiotic behavior of inversions	16

2. Malignancy and acquired chromosome abnormalities.
B.H. Czepulkowski — **17**

Classification of disease	17

Cytogenetic changes in malignancy	18

Tables

FAB classification of AML	19
Chromosome abnormalities in AML	20
FAB classification of MDS excluding CMML	22
Chromosome changes in MDS	22
Classification of MPD	23
Chromosome changes in MPD	24
Chromosome changes in ALL	24
Common chromosome changes in transformed CGL	25
FAB classification of ALL	26
Chromosome abnormalities in lymphomas	27
Chromosome changes in chronic lymphoproliferative disorders	28
Translocations and oncogenes	29
Common chromosome changes in solid tumors	31
Symptoms of diseases encountered	34

bright field microscope	55
Engravings on an objective lens	56
Annular rings observed by phase contrast microscopy	58
Arrangement of lens elements and light paths	61
Light source and filters in a reflected light (epi-) fluorescence microscope	63
Filters and light paths in a reflected light fluorescence microscope	62
Staining methods	66
Fluorochromes used in cytogenetics	73
Photographic equipment and materials	74

5. Culturing cells for chromosome preparation. D.E. Rooney and B.H. Czepulkowski **76**

Tables

Media for cell and tissue culture for cytogenetic study	78

3. The relationship between cytogenetic and molecular genetic studies. J. Waters and F. MacDonald — **40**

Table
Cytogenetic and molecular genetic approaches to disease — 42

4. Analyzing chromosomes: staining, banding and microscopy. A.J. Monk — **54**
Microscope adjustment for bright field observation — 55
Microscope adjustment for phase contrast observation — 57
Objective lens markings — 59
The fluorescence microscope — 61
Computerized imaging systems — 63
Photography — 65

Figures and Tables
Light source, lens elements and controls on the

Reagents used for routine cell and tissue culture — 79
Culture conditions for expression of specific chromosome abnormalities and methods of analysis — 81
Principal types of cultured amniotic fluid cells and their properties — 83
Normal hematological values — 84
Bone marrow sample seeding volumes according to white cell count — 85
Guidelines for setting up bone marrow samples — 86
Reasons for referral of bone marrow samples with potential diagnoses — 87

6. Prenatal diagnosis of chromosome abnormality. D.E. Rooney — **89**
Screening methods — 89
Methods of obtaining a fetal karyotype — 91
Fetal measurement — 91

Tables

Rates of chromosome abnormalities in live births by single-year interval	92
Maternal-age-specific risks for trisomy 21	94
Maternal-age-specific risks for trisomies 18 and 13	95
Maternal-age-specific risks for chromosomal abnormalities at chorionic villus sampling (excluding trisomy 21)	94
Maternal serum screening for Down syndrome: using age, AFP, uE_3 and hCG	95
Serum screening for Down syndrome: DR, FPR and OAPR using maternal age, AFP, uE_3 and hCG according to risk cut-off, method of gestational age and maternal weight	96
Chromosome abnormalities associated with fetal structural abnormalities	97
Definitions of mosaicism in CVS and amniotic fluid	97
CVS level III mosaicism	97

Chromosome instability syndromes	106

Figure and Tables

Chromosome damage identified by block staining	104
Clinical features and patterns of inheritance	109
Spontaneous and induced cytogenetic changes in chromosome instability syndromes	110

8. Health and safety data. R.T. Howell and S.H. Roberts **111**

General data	111
Pathological hazards	113
Decontamination and waste disposal	113
Chemical hazards	115
Publications concerning statutory instruments, Health and Safety Executive guidelines and directives	117

Tables

Containment of pathological hazards	119

Potential significance and value of chromosome results from chorionic villi	98
Percentage mosaicism excluded for specified number of cells evaluated	98
Assessment of gestational age from crown–rump length	99
Predicted fetal measurements at specific menstrual age	100

7. The role of cytogenetics in the investigation of mutagen exposure and chromosome instability.
E.J. Tawn and D. Holdsworth **102**

Description of cytogenetic endpoints	103
Applications	103
Activities of disinfectants	121
Chemicals assigned inhalation exposure limits	120
Risk phrases for hazardous chemicals	120
Cytotoxic chemicals	121

9. Manufacturers and suppliers **123**

References **132**

Further reading **140**

Appendix **142**
Ideogram of G-banded human karyotype 142

Index **144**

CONTRIBUTORS

I. Cross
Department of Human Genetics, University of Newcastle upon Tyne, 19 Claremont Place, Newcastle upon Tyne NE2 4AA, UK

B.H. Czepulkowski
Cytogenetic Unit, Department of Haematology, King's College Hospital, Denmark Hill, London SE5 9RS, UK

D. Holdsworth
Geoffrey Schofield Cytogenetics Laboratory, BNF plc, Sellafield, Seascale, Cumbria CA20 1PG, UK

R.T. Howell
SW Regional Cytogenetics Centre, Southmead Hospital, Westbury on Trym, Bristol BS10 5NB, UK

S.H. Roberts
Regional Cytogenetics Unit, Institute of Medical Genetics for Wales, University Hospital of Wales, Heath Park, Cardiff CF4 4XW, UK

D.E. Rooney
Cytogenetic Services, 48 Wimpole Street, London W1M 7DG, UK

E.J. Tawn
Geoffrey Schofield Cytogenetics Laboratory, BNF plc, Sellafield, Seascale, Cumbria CA20 1PG, UK

J.J. Waters
Regional Cytogenetics Laboratory, East Birmingham Hospital, Bordesley Green East, Birmingham B9 5ST, UK

F. MacDonald
Regional Cytogenetics Laboratory, East Birmingham Hospital, Bordesley Green East, Birmingham B9 5ST, UK

A.J. Monk
Cytogenetics Department, City Hospital, Hucknall Road, Nottingham NG5 1PB, UK

J. Wolstenholme
Department of Human Genetics, University of Newcastle upon Tyne, 19 Claremont Place, Newcastle upon Tyne NE2 4AA, UK

ABBREVIATIONS

AC	amniocentesis
AFP	alphafetoprotein
ALL	acute lymphoblastic leukemia
AML	acute myeloid leukemia
ANLL	acute nonlymphocytic leukemia
APML	acute promyelocytic leukemia
ASD	atrial septal defect
AT	ataxia telangiectasia
AT-rich	adenine–thymine-rich
ATL	adult T-cell leukemia/lymphoma
BL	Burkitt's lymphoma
BM	bone marrow
BPD	biparietal diameter
BrdU	bromodeoxyuridine
CG-rich	cytosine–guanine-rich
CGH	comparative genomic hybridization
CGL	chronic granulocytic leukemia
CLL	chronic lymphocytic leukemia

ESR	erythrocyte sedimentation rate
ET	essential thrombocythemia
FAB	French–American–British (classification)
FdU	fluorodeoxyuridine
FISH	fluorescence *in situ* hybridization
FITC	fluorescein isothiocyanate
FPR	false positive rate
HBSS	Hank's balanced salt solution
hCG	human chorionic gonadotropin
HCL	hairy-cell leukemia
HD	Hodgkin's disease
HSE	Health and Safety Executive
HSR	homogeneously staining regions
ICF	immunodeficiency, centromeric heterochromatin instability and facial anomalies
LB	live birth
LCL	lymphoblastoid cell line

CML	chronic myeloid leukemia	LGLL	large granular lymphocytic leukemia
CMML	chronic myelomonocytic leukemia	MA	maternal age
CRL	crown–rump length	MCV	mean corpuscular volume
CS	Cockayne's syndrome	MDS	myelodysplastic syndromes
CVS	chorionic villus sampling	MEL	maximum exposure limit
DA-DAPI	distamycin A–DAPI	MEM	minimal essential medium
DAPI	4,6-diamidino-2-phenylindole	MF	myelofibrosis
DIC	disseminated intravascular coagulation	MIC	morphologic, immunologic and cytogenetic
dmin	double minutes	MM	multiple myeloma
DMSO	dimethylsulfoxide	MPD	myeloproliferative disease
DNA	deoxyribonucleic acid	MTX	methotrexate
DPX	a microscopic mountant	NAP	neutrophil alkaline phosphatase
DR	detection rate	NHL	non-Hodgkin's lymphoma
EA	early amniocentesis	NOR	nucleolar organizer region
EB	Epstein–Barr	OAPR	odds of being affected given a positive screening result (for Down syndrome)
EBSS	Eagle's balanced salt solution		
EDD	estimated date of delivery	PB	peripheral blood
EDTA	ethylenediaminetetraacetic acid	PCL	plasma-cell leukemia
EGFR	epidermal growth factor receptor	PCR	polymerase chain reaction

PCV	packed cell volume
PFGE	pulsed field gel electrophoresis
PGL	persistent generalized lymphadenopathy
PHA	phytohemagglutinin
PLL	prolymphocytic leukemia
PMA	phorbol 12-myristate 13-acetate ($\equiv$TPA)
PRV	polycythemia rubra vera
PWM	pokeweed mitogen
RA	refractory anemia
RAEB	refractory anemia with excess blasts
RAEBT	refractory anemia in transformation
RARS	refractory anemia with ringed sideroblasts
RAS	refractory anemia with sideroblasts
RFLP	restriction fragment length polymorphism

RT	reverse transcription
SCE	sister chromatid exchange
TPA	12-O-tetradecanoyl phorbol-13-acetate ($\equiv$PMA)
TTD	trichothiodystrophy
TWA	time-weighted average
uE$_3$	unconjugated estriol
UV	ultraviolet
VSD	ventricular septal defect
WCC	white cell count
WM	Waldenström's macroglobulinemia
XP	xeroderma pigmentosum
YAC	yeast artificial chromosome

PREFACE

The concept of 'Essential Data' in human cytogenetics has proved quite a challenge to us and our authors: in a book of this size, what sort of data can one regard as essential? After much discussion with the authors and other helpful cytogeneticists the picture of this book became clearer in our 'mind's eye'. So we set out to extract from the wealth of literature those facts and figures that are useful to have to hand in one place. In some cases, this includes data which is not, strictly speaking, cytogenetic, but which the cytogeneticist needs to consult every so often (the obstetric data, for instance). We hope that this book will be regarded as a useful complement to the more technique-orientated books already in existence.

This book is dedicated to Bill Lockley and his group in Dundee.

D.E. Rooney
B.H. Czepulkowski

Chapter 1 CONSTITUTIONAL CHROMOSOME ABNORMALITIES –

J. Wolstenholme and I.E. Cross

More than 99% of individuals in newborn surveys have a normal 46,XX or 46,XY karyotype. The basic features of mitotic and meiotic chromosomes in such normal individuals are given in *Table 1*. *Tables 2* and *3* provide incidences of trisomies, common sex chromosome anomalies and other numerical abnormalities in different referral categories during pregnancy through to delivery, with specific additional data for Turner's syndrome in *Table 4*. The meiotic origin of some of these anomalies has been established (*Table 5*).

Population incidences of structural abnormalities appear in *Tables 6* and *7*, with trivalent segregation terminology for Robertsonian translocations being given in *Table 8*. The frequency of observation of balanced and unbalanced products of these Robertsonian translocations during prenatal diagnosis depends on the gestation time at sampling, with the more unbalanced (nonviable) forms being detected only in the first trimester (*Table 9*). *Table 9* also gives some indication of the relative frequency of each translocation, although there will be some ascertainment bias towards those causing clinical problems. Terminology for the common unbalanced forms of reciprocal translocations and inversions is given in *Table 10* and *Figures 1* and *2*. The segregation products which are encountered in the laboratory will depend on a combination of meiotic quadrivalent behavior and the viability of unbalanced rearrangements *in utero*, with highly unbalanced forms being lost early in pregnancy. With the exception of the well characterized t(11;22)(q23;q11), all need to be considered individually, so no equivalent of *Table 9* can be produced.

Human Cytogenetics

Table 1. The human karyotype

Chromosome	Morphology [1]		Chiasmata [1]		Polymorphisms	Fragile sites [2][e]	Other information
	Group	Size[c]	Mean	%			
1	Metacentric			2–5	Centromeric C-band: small, large, partial inversions; some large variants may appear as two distinct bands	p36, p32, p31.2, p31, p22, p21.2, q12[a], q21, q25.1, q31, q42[a], q44	Centromere DA-DAPI +ve, variable
	A	8.44%	3.90	7.59			
2	Metacentric			2–5	Inv(p11q13)	p24.2, p16.2, p13, $q11.2^f$, $q13^f$, p21.3, $q22.3^f$, q31, q32.1, q33, q37.3	
	A	8.02%	3.62	7.11			
3	Metacentric			2–4	Centromeric C/Q-band intensity	p24.2, p14.2, q27, q25	
	A	6.83%	2.92	5.81			
4	Submetacentric			2–3	Centromeric C/Q-band intensity	p16.1, p15, q12[b], $q27''$, q31.1	
	B	6.30%	2.79	5.48			
5	Submetacentric			2–4		p14, p13[b], q15, q15[b], q21, q31.1	
	B	6.08%	2.85	5.53			
6	Submetacentric			2–4		p25.1, $p23^f$, p22.2, q13[b], q15, q21, q26	
	C + X	5.90%	2.67	5.04			
7	Submetacentric			2–4		p22, p14.2, p13, $p11.2^f$, q11, q21.2, q22, q31.2, q32.3, q36	
	C + X	5.36%	2.74	5.21			
8	Submetacentric			2–3		$q13^d$, q22.1, $q22.3^f$, q24.1, $q24.1^d$, q24.3	
	C + X	4.93%	2.64	4.99			
9	Submetacentric			2–3	Centromeric C-band: small; inverted (1% of	p21[b], $p21^f$, q12[a], q22.1, q32, $q32^f$	Centromere DA-DAPI +ve, variable
	C + X	4.80%	2.41	4.73			

No.	Morphology / Group	%	min	max	Notes	Bands	Other
10	Submetacentric / C+X	4.59%	2.50	4.96	population); large – can appear to contain dark G-band	q21[b], q22.1, *q23.3* or *q24.2*[f], q25.2, *q25.2*[b], q26.1	
11	Submetacentric / C+X	4.61%	2.21	4.31		p15.1, *p15.1*[d], p14.2, p13, q13, *q13.3*[f], q14.2, q23.3, *q23.3*[f]	
12	Submetacentric / C+X	4.66%	2.71	5.22		*q13.1*[f], q21.3, q24, *q24.13*[f], *q24.2*[b]	
13	Acrocentric / D	3.74%	1.85	3.63	Considerable variation of centromere, proximal short arm, stalks (NOR stain) and satellites	q13.2, q21[b], q21.2, q32	
14	Acrocentric / D	3.56%	1.88	3.66	As chromosome 13	q23, q24.1	
15	Acrocentric / D	3.46%	2.05	4.03	As chromosome 13; very large p-arm variants are known	q22	Proximal p-arm DA-DAPI +ve
16	Metacentric / E	3.36%	2.16	4.24	Centromeric C-band: large; small	*p13.11*[f], *p12.1*[d], q22.1, *q22.1*[d], q23.2	Centromere DA-DAPI +ve, variable
17	Submetacentric / E	3.25%	2.13	4.14		*p12*[d], q23.1	
18	Submetacentric / E	2.93%	1.92	3.73		q12.2, q21.3	

Continued

Table 1. The human karyotype, *continued*

Chromosome	Morphology [1]		Chiasmata [1]		Polymorphisms	Fragile sites [2][e]	Other information
	Group	Size[c]	Mean	%			
19	Metacentric			1–2	Centromeric G/C-band may appear on both p and q arms, with pale discontinuity between	*p13*[f], q13[a]	
	F	2.67%	1.94	3.82			
20	Metacentric			1–2		p12.2, *p11.23*[f]	
	F	2.56%	2.00	3.98			
21	Acrocentric			1–2	As chromosome 13		
	G + Y	1.90%	1.05	2.08			
22	Acrocentric			1–2	As chromosome 13; large variants almost q-arm size are known	q12.2, *q13*[f]	
	G + Y	2.04%	1.22	2.39			
X	Submetacentric			—		p22.31, q22.1, q27.2, *q27.3*[f] (FRAXA), *q27.3*[f] (FRAXE)	
	C + X	5.12%	—	—			
Y	Submetacentric			—	q12 G/C-band: small, can be apparently missing; large, can produce E- or even D-size chromosome		q12 DA-DAPI +ve
	G + Y	2.15%	—	—			

[c]Percentage of haploid autosomal length.
[e]Common, *rare*; aphidicolin type except: [a]5-azacytidine type; [b]bromodeoxyuridine (BrdU) type; [d]distamycin-A type; [f]folic acid type; [u]unclassified.

Table 2. Incidence of constitutional abnormalities in spontaneous losses and chorionic villus sampling (CVS)

Chromosome abnormality	Spontaneous losses [3] 3300 cases up to 28 weeks[a]		CVS – combined data from US [4] and UK [5] collaborative studies, 18 851 cases			
	Nonmosaic	Mosaic[b]	Nonmosaic	Confirmed	Mosaic	Confirmed
(a) Trisomies						
1	—	—	—	—	—	—
2	34	3	—	—	11	0/2
3	6	1	1	0/1	9	0/5
4	15	—	—	—	—	—
5	5	—	—	—	1	0/1
6	5	1	—	—	1	—
7	27	2	2	0/2	15	0/8
8	23	3	—	—	9	0/8
9	18	—	1	—	5	0/3
10	11	3	—	—	1	0/1
11	—	—	—	—	1	0/1
12	2	—	—	—	9	0/8
13	53	—	27	14/14	6	1/4
14	32	1	2	—	1	0/1
15	52	1	1	—	4	0/3
16	202	16	6	1/4	1	0/1
17	4	1	—	—	—	—
18	23	1	69	33/33	10	2/8
19	—	—	—	—	1	0/1
20	18	5	—	—	5	0/2
21	54	1	209	103/103	11	3/9
22	55	2	8	0/2	2	0/2

Continued

Table 2. Incidence of constitutional abnormalities in spontaneous losses and CVS, *continued*

Chromosome abnormality	Spontaneous losses [3] 3300 cases up to 28 weeks[a]		CVS – combined data from US [4] and UK [5] collaborative studies, 18 851 cases			
	Nonmosaic	Mosaic[b]	Nonmosaic	Confirmed	Mosaic	Confirmed
(b) Others						
45,X	201	15 X/XX	28	15/15[c]	25 X/XX	3/11
					9 X/XY	1/7
47,XXY	8	—	23	12/12	1	—
47,XYY	—	—	12	7/7	—	—
47,XXX	—	—	8	4/4	3	1/2
Other sex chrom.	—	—	1	—	11	2/6
+mar	—	—	2	1/2[d]	13	6/9
Triploid	185	—	15	6/6	1[e]	—
Tetraploid	65	—	3	0/3	13	0/8
Double trisomy	23	12[f]	3	0/1	3	0/2

[a]Most chromosome abnormalities lost prior to 14 weeks' gestation.
[b]Details of only 41/60 mosaic autosomal trisomy cases available [6].
[c]One case fetal 45,X/46,XX.
[d]Fetus 46,XX/46,XX, +mar.
[e]Cultured cells only, ? maternal cell contamination.
[f]Mosaic trisomy/double trisomy.

Table 3. Incidence of constitutional abnormalities at amniocentesis and birth

Chromosome abnormality	Amniocentesis [7] in 52 965 analyses European Study[a] (+ mosaics)[b]	Mosaic [8] US study		Pseudomosaic [8] US study		Birth	Comments
		In 62 279 analyses + reports[c]	Confirmed	Multiple cells in 48 442 analyses	Single cells in 30 754 analyses[d]		
(a) Trisomies							
1	—	—	—	—	9	None	
2	1	—	—	28	73	None?	One unconfirmed mosaic [9]
3	—	—	—	—	12	Very rare	Mosaic only [10]
4	—	—	—	1	7	None	
5	—	—	—	3	18	None	
6	—	—	—	1	11	None	
7	—	1	0/1	10	14	Very rare	Mosaic only [11]
8	—	5	2/4	3	14	Rare	Probably all mosaic
9	—	6	2/5	5	12	Very rare	Usually mosaic [12,13]
10	—	—	—	4	14	Very rare	Mosaic only [14]
11	—	1	0/1	2	14	None	
12	—	1	0/1	1	16	Very rare	Mosaic only [15]
13	39(+1)	2	1/1	3	7	1 in 5000	15% are mosaics
14	1	1	—	2	10	Very rare	Mosaic only [16]
15	—	1	1/1	2	8	Very rare	[17,18]
16	—	1	—	—	14	Very rare	Mosaic only [19]
17	—	1	0/1	5	8	Very rare	Mosaic only [20]

Continued

8 *Human Cytogenetics*

Table 3. Incidence of constitutional abnormalities at amniocentesis and birth, *continued*

Chromosome abnormality	Amniocentesis [7] in 52 965 analyses European Study[a] (+ mosaics)[b]	Mosaic [8] US study		Pseudomosaic [8] US study		Birth	Comments
		In 62 279 analyses + reports[c]	Confirmed	Multiple cells in 48 442 analyses	Single cells in 30 754 analyses[d]		
18	121(+2)	4	4/4	1	12	1 in 3000	10% are mosaics
19	—	—	—	1	9	Very rare	[21]
20	(+3)	20	6/16	5[e]	21	Rare	Mosaic only, blood usually negative [22]
21	613(+4)	21	15/17	2	15	1 in 800	3% are mosaics
22	(+3)	1	0/1	1	12	Very rare	Nonmosaic recorded [23]
(b) Others							
45,X	24 (+5 X/XX) (+2 X/XY)	21 X/XX 17 X/XY	13/14 8/13	14 X/XX[e] 14 X/XY	NA[f]	1 in 5000[g] for 45,X	1 in 5000 mosaic or other X abnormality[h], see *Table 4*
47, XXY	87 (+3)	13	10/11	NA	NA	1 in 1000[g]	10–20% are mosaic
47, XYY	18	1	—	2	8	1 in 1000[g]	~10% are mosaic
47, XXX	65 (+1)	6	5/5	NA	NA	1 in 1000[g]	~10% are mosaic
Other sex chrom.	16 (+7)	15	11/11	NA	NA		
+ mar	31	21	13/16	NA	NA	1 in 2500	~20% are mosaic
Triploid	3	2	2/2	—	—		May reach term but very low viability in neonatal period

Tetraploid	—	—	—	—		
Double trisomy	—	—	—	—	Not viable	Viable trisomy 21 with sex chromosome abnormality recorded

[a] Maternal age referrals only.
[b] No data on confirmation of mosaics.
[c] 159 mosaics from series of 62 279 plus 29 from other documented sources.
[d] Very variable rates from individual laboratories. Other data [24] indicate gross under-reporting of single trisomic cells and most pseudomosaics for structural abnormalities.
[e] One case confirmed (trisomy 20 case in repeat amniocentesis only).
[f] Detailed data not available in study.
[g] Incidence in males or females as appropriate.
[h] Based on rates of clinical ascertainment of Turner's phenotype not birth incidence; underestimation of karyotypes with minimal phenotypic effects.

Table 4. Karyotype variation in Turner syndrome

Karyotype	Incidence in 110 cases of Turner syndrome with a chromosome abnormality [25]	%
45,X	64	58.2
45,X/46,XX or 47,XXX	11	10
46,X,i(Xq) (including mosaics)	21	19.1
45,X/46,X,r(X)	6	5.5
45,X/46,XY or 47,XYY	6	5.5
46,X,Xp-	1	0.9
46,X,t(X;autosome)	1	0.9

Reproduced from ref. 25 with permission from Springer Verlag.

Table 5. Parental origin of common constitutional chromosome abnormalities

Abnormality	Origin[a]
Trisomy 13	Maternal I, 33%; maternal II, 13%; maternal either, 10%. paternal I, 7%; paternal II, 3%; uninformative, 33% [26]
Trisomy 16	All maternal meiosis I [27]
Trisomy 18	Extra maternal chromosome, 92%; extra paternal chromosome, 8% [28,29]; up to 21% are postzygotic errors [29]
Trisomy 21	Maternal I, 64%; maternal II, 19%; maternal either, 11%; paternal I, 1%; paternal II, 3.5%; unknown, 1.5% [30]
45,X	Maternal X present, 72%; paternal X present, 28% [31]
45,X/46,X,i(Xq)	Paternal, 52%; maternal, 48% [32]
46,XX/46,X,i(Xq)	Paternal, 23%; maternal, 77% [32]
XXX	Maternal I, 54%; maternal II, 21%; maternal either, 18%; paternal, 7% [33]
XXY	Maternal, 50%; paternal, 50% [34]
Triploidy	Dispermy, 39%; dispermy or paternal I, 15%; maternal I, 8%; maternal II, 12%; uninformative, 26% [35,36]

[a]Trisomies are usually assigned to meiosis I or II nondisjunction; an alternative interpretation of apparent meiosis I errors, particularly for trisomy 16, has been proposed [37]; some apparent meiosis II events may be postzygotic errors [29,30].

Table 6. Incidence of structural chromosome abnormalities

Type of abnormality	Incidence/1000 in amniotic fluids referred for advanced maternal age [7]	Incidence/1000 in live births
Extra marker chromosome	0.59	0.63[a]
Balanced reciprocal translocation	1.21	2.00[a]
Balanced Robertsonian translocation	0.74	1.35[a]
Unbalanced Robertsonian translocation	0.13	0.07[b]
Inversion[c]	0.36	0.55[a]
Other unbalanced structural abnormalities	0.26	0.19[a]
inv(9)(p11q13)	—	8.66[a]

[a] Data pooled from five studies [38–42]; [b] Data from ref. 42 only; [c] Excluding inv(9)(p11q13).

Table 7. Incidence of rarer structural chromosome abnormalities

Type of abnormality	Incidence/1000 in live births	Useful references
Insertion	0.2	43
Deletion	0.11[a,b]	
Ring chromosome	0.015[a]	44
Derivative chromosome	0.03[a]	

[a] Data from ref. 45.
[b] Data derived from series published between 1969 and 1980. This will be an underestimate of the incidence of deletions as it will not include microdeletions recognized subsequent to those dates.

Table 8. Robertsonian translocations – trivalent segregation, e.g. t(14q21q)

	Alternate segregation	Adjacent segregation		3:0 Segregation
Gametes and outcomes	14, 21 Normal	21, t(14q21q) Trisomy 21	14, t(14q21q) Trisomy 14	14, 21, t(14q21q) Trisomy 14 and 21
	t(14q21q) Balanced	14 Monosomy 21	21 Monosomy 14	— Monosomy 14 and 21

Alternate
normal/balanced

Adjacent[a]
trisomy 21/monosomy 21

Adjacent[a]
trisomy 14 /monosomy 14

[a]Without knowledge of which centromere is present on the translocation chromosome, adjacent 1 and adjacent 2 segregations cannot be distinguished.

Table 9. Segregation patterns of Robertsonian translocations at amniocentesis and CVS

Translocation	Amniocentesis [46]: offspring				CVS [47]: offspring			
	Number	Normal	Balanced	Unbalanced[a]	Number	Normal	Balanced	Unbalanced[a]
13q14q mat	157	69	88	—	20	7	11	2[b,c]
13q14q pat	73	27	46	—	5	2	3	—
13q15q	15	4	11	—	—	—	—	—
14q15q	6	4	2	—	—	—	—	—
13q21q	31	17	12	2[d]	1	—	1	—
13q22q	3	—	2	1[b]	1	—	1	—
14q21q mat	137	48	68	21[d]	12	2	7	3[d,e]
14q21q pat	51	20	31	—	4	3	1	—
14q22q	3	—	3	—	1	1	—	—
15q21q	14	7	6	1[d]	—	—	—	—
15q22q	5	2	3	—	—	—	—	—
21q22q	22	11	8	3[d]	1	—	1	—
Total	517	209	280	28	45	15	25	5

[a] All maternally derived.
[b] Trisomy 13.
[c] Trisomy 14.
[d] Trisomy 21.
[e] One case 47,XY,-14,+t(14q21q),+t(14q21q).

14 Human Cytogenetics

Table 10. Reciprocal translocations – quadrivalent segregation

	Alternate segregation	Adjacent 1 segregation	Adjacent 2 segregation	3:1 Segregation				4:0 Segregation
Gametes and outcomes	A B Normal	A B′ Unbalanced	A A′ Unbalanced	A A′ B′ Interchange trisomy[a]	A′ B′ B Interchange trisomy	A B B′ Tertiary trisomy	A A′ B Tertiary trisomy[a]	A A′ B B′ Double trisomy[a]
	A′ B′ Balanced	A′ B Unbalanced	B B′ Unbalanced	B Interchange monosomy[a]	A Interchange monosomy	A′ Tertiary monosomy	B′ Tertiary monosomy[a]	— Double monosomy[a]
Predisposing factors	—	Short translocated segments	Short centric segments	One chromosome in quadrivalent relatively small e.g. t(11;22)(q23;q11) [48]				—

[a]This segregation is not shown in *Figure 1.*

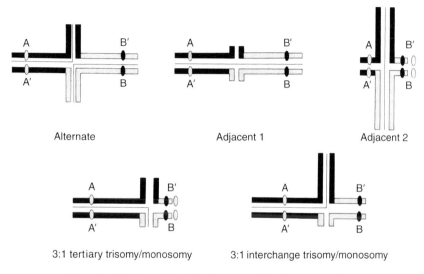

Figure 1. Reciprocal translocations – quadrivalent segregation.

Human Cytogenetics

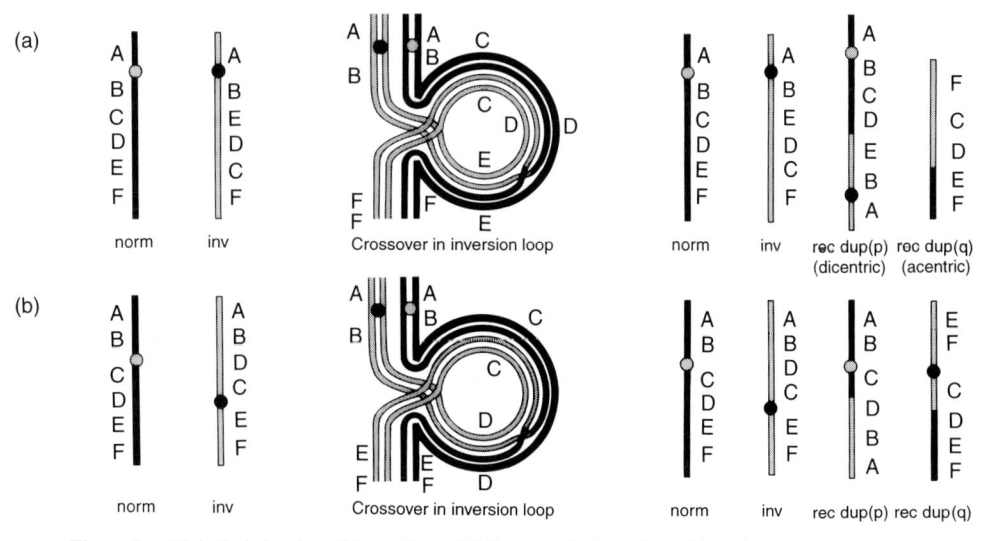

Figure 2. Meiotic behavior of inversions. (a) Paracentric inversion; (b) pericentric inversion.

Chapter 2 MALIGNANCY AND ACQUIRED CHROMOSOME ABNORMALITIES – B.H. Czepulkowski

Chromosome studies on malignant cells in both man and mouse have indicated for many years that the abnormal changes observed are an integral aspect of the evolution of such disease. Originally, studies were performed on highly malignant tumors which proved difficult to interpret as many changes were observed in these. Consistent abnormalities were noted, but they existed with a number of other changes and, thus, the primary change could not be identified with any degree of accuracy. It was the discovery of a small marker chromosome by Nowell and Hungerford in 1960, which really increased the interest in cancer cytogenetics. The marker (Philadelphia chromosome) was observed in patients with chronic granulocytic leukemia (CGL). Following the introduction of banding in the 1970s, the small marker was re-evaluated as a translocation between chromosomes 9 and 22. This led to a whole host of new associations which has expanded as time has gone by, and at the present day is still growing. The leukemias and other hematological disorders are still the diseases which are most studied, as bone marrow and blood are most easily accessible. However, some centers specialize in particular areas, such as tumors, lymphomas or breakage syndromes and chromosome damage.

1 Classification of disease

Tables 1–13 give a brief description of the classification of various hematological malignancies and lymphomas. These are used by clinicians to facilitate the appropriate treatment for various types of disease. Most require immunological techniques in addition to morphology and cytogenetics. Using prognostic information gained from cytogenetic findings, patient management is greatly enhanced. *Table 14* gives symptoms and hematological findings of diseases encountered.

2 Cytogenetic changes in malignancy

Tables 1–11 and *13* give lists of a number of changes which have been associated with types of disease in hematological malignancies, chronic lymphoproliferative disorders, lymphomas and solid tumors. Where the incidence of these changes is known it is given in the body of the table. Some changes have important prognostic implications and these are also shown. *Table 12* shows a number of translocations associated with disease with their respective involved genes at breakpoint sites. Further lists of oncogenes involved in constitutional and acquired abnormalities are given in Chapter 4. Chapter 5 gives details of the methods utilized in preparing bone marrow or blood in order to analyze the chromosome changes shown here.

Table 1. FAB classification of AML

FAB type	Incidence (%)	Description
M1	15–20	Myeloblastic without maturation
M2	30	Myeloblastic with maturation
M3	5–10	Promyelocytic (hypergranular)
M3 variant	5–10	Promyelocytic (hypo- or micro-granular)
M4	15–20	Myelomonocytic
M4Eo	15–20	M4 with eosinophilia
M5a	15	Monoblastic
M5b	15	Promonocytic–monocytic
M6	3–4	Erythroblastic (< 30% blasts if > 50% erythroblasts)
M7	2–4	Megakaryoblastic
M0	< 1	Myeloblastic with minimal differentiation

The FAB classification was published in 1976, compiled by a 2 year study via a cooperative group as an acceptable system of classification of acute myeloid leukemia (AML) [1]. MIC (morphologic, immunologic and cytogenetic) classifications of AML are brought together as suggested by the MIC group [2]. AML incidence: 1 in 100 000 per year in children; 1–10 in 100 000 per year in adults.

Human Cytogenetics

Table 2. Chromosome abnormalities in AML

Chromosome localization	Type of rearrangement and incidence[a]	Association with disease and prognosis[a]
1p11	t(1;7)(p11;p11)	Secondary AML mostly M4; poor prognosis
3q21 and q26	ins(3;3)(q26;q21q26) inv(3)(q21q26) t(3;3)(q21;q26) t(3;5)(q21–25;q31–35) <1%	Abnormal megakaryocytes and thrombocytosis, fibrosis; poor prognosis, patients elderly
4	Trisomy 4 <1%	M2 and M4
5	Monosomy 5	Secondary AML; poor prognosis
5q	del(5q)	Secondary AML; poor prognosis
6p23	t(6;9)(p23;q34) t(6;11)(q27;q23) <1%	M2 and M4; poor prognosis
7	Monosomy 7	Secondary AML; poor prognosis
7p15	t(7;11)(p15;p15)	M2 with MDS features
7q	del(7q)	Secondary AML; poor prognosis
8	Trisomy 8	Myeloid disease
8p11	t(8;16)(p11;p13) <1%	M5 or M5a, phagocytosis, mainly infants and children
8q22	t(8;21)(q22;q22) 9%	M2 with Auer rods, eosinophilia; intermediate prognosis, in younger patients
9p21	t(9;11)(p21;q23)	M5 mostly M5a; intermediate prognosis
9q13 or q22	del(9) <1% secondary association with t(8;21)	Agranular blasts
9q34	t(9;22)(q34;q11) <1%	M1 and M2
10p14	t(10;11)(p14;q13)	M4, M5, M0, M0Baso

11	Trisomy 11	Multiple FAB types; MDS features
11q23	del/t(11q) 4%	M4, M5, mostly M5a, mostly children and infants; intermediate prognosis
12p11	del/t(12p)	Secondary AML, M2, M4Baso; poor prognosis
15q22	t(15;17)(q22;q11) 9%	M3 and M3v often with +8 or i(17q)
16p13 and q22	t(16;16)(p13;q22)	M4Eo; good prognosis, younger patients, secondary changes +8, +22 and del(7q)
	inv(16)(p13q22)	
	del(16)(q22q24)	del(16) more often found in MDS
17	i(17q)	Multiple FAB types MDS features
20q	del(20)(q11q13.1)	M6
	del(20)(q11q13.3)	
21	Trisomy 21	Multiple FAB types MDS features
22	Trisomy 22	M4 MDS features (look out for inv(16) with this change!)
Y	Monosomy Y	Age-related phenomenon

[a]If known.

MDS, myelodysplastic syndromes.

Table 3. FAB classification of MDS excluding CMML

Disease and incidence within MDS	% blasts in marrow	Other features
Refractory anemia (RA) 30–40%	< 5 Anemia in blood, blasts ≤1%, monocytes ≤1 x $10^9 l^{-1}$	< 15% ringed sideroblasts in erythroblasts
Refractory anemia with sideroblasts (RAS) 15–25%	< 5 Anemia in blood, blasts ≤1%, monocytes ≤1 x $10^9 l^{-1}$	> 15% ringed sideroblasts in erythroblasts
Refractory anemia with excess blasts (RAEB) 15–25%	5–20	—
Refractory anemia in transformation (RAEBT) 15–25%	21–29[a] In blood blasts ≥5%	Also RAEBT if Auer rods present, irrespective of blast count

Classification of myelodysplastic syndromes (MDS) with five categories being recognized by the FAB group (this includes chronic myelomonocytic leukemia (CMML)) [3–5]. Cytogenetics and immunophenotyping can be incorporated into the classification as proposed by the MIC group [6].
[a]A blast count of ≥30% is a diagnostic criterion for AML.
Incidence of MDS 0.75 per 1000 per year in those patients over 60 years.

Table 4. Chromosome changes in MDS

Abnormality	Incidence (%)
del(5)	27
Monosomy 7	15
del(7)	4
Trisomy 8	19
del(11) (q14) or (q23) or t	7

del(13q)	2
del(20)(q11q13)	5
−Y	5
i(17q)	2
t or del(12)(p11–p13)	5
3q26 abnormalities	1
t(1;7)(p11;p11)	2
t(1;3)(p36;q21)	1

Changes which specifically transform to AML
t(2;11)(p21;q23)
t(11;21)(q24;q11.2)
t(1;3)(p36;q21)
t(6;9)(p25;q34)
t(1;7)(p11;q11)

Less common changes
del(1p)
−15, +der(15)t(1;15)(q11;p11)
del(17p)
−20
t(21q22)
del(21q)
−22

Table 5. Classification of myeloproliferative disorders (MPD)

% with abnormality	Disease	Major proliferative component
90	Chronic myeloid leukemia (CML) Chronic granulocytic leukemia (CGL)	Myeloid activity predominates; 1 in 100 000 per year, 60 years median age, high white cell count, granulocytic and megakaryocytic hyperplasia
30	Chronic myelomonocytic leukemia (CMML)	
15	Polycythemia rubra vera (PRV)	Red cell activity predominates, erythrocytosis
5	Essential thrombocythemia (ET)	Platelet activity predominates, thromboembolitic and hemorrhagic phenomena
50	Myelofibrosis (MF), myeloid metaplasia	Reactive marrow fibrosis predominates

Table 6. Chromosome changes in MPD

Abnormality	Incidence
t(9;22)(q34;q11) (CML and CGL)	Most CMLs
+1q	
del(5q)	
−7,der(7)t(1;7)(p11;p11)	
Monosomy 7	
del(7q)	
Trisomy 8	
Trisomy 9	
del(11)(q23)	
del(12p)	
del(13q)	
del(20)(q11q13)	

Table 8. Chromosome changes in ALL

Abnormality	Association (if any)
t(8;14)(q24;q32)	L3; poor prognosis
t(8;22)(q24;q11)	L3; poor prognosis
t(2;8)(p12;q24)	L3; poor prognosis
Duplications of 1q	L3
t(1;19)(q23;p13)	L1 pre-B-cell ALL
t(1;11)(p32;q23)	L1 pre-B-cell ALL
t(7;12)(q11;p12)	B lineage ALL
dic(9;12)(p11;p12)	B lineage ALL
l(1,14)(p32;q11)	T lineage ALL
t(8;14)(q24;q11)	T lineage ALL
t(10;14)(q24;q11)	T lineage ALL
t(11;14)(p13;q11)	T lineage ALL
t(7;14)(q35;q11)	T lineage ALL
inv(14)(q11q32)	Adult T-cell leukemia
i(6p)	
del(6q)	Common T/B lineage; intermediate prognosis
dic(7;9)(p11;p11)	
i(7q)	
del(9)(p21)	L1/L2, T/B lineage
t(9;22)(q34;q11)	L1/L2; poor prognosis, immature B cell
t(4;11)(q21;q23)	Poor prognosis; immature B cell

Table 7. Common chromosome changes in transformed CGL (major routes – 70% of transformed CGLs)

Additional change	Frequency (%)
Trisomy 8	8
i(17q) and abnormalities of 17p	12
+der(22) Philadelphia chromosome	14
Trisomy 19	1
+der(22), +8	9
+8, i(17q)	9
+19, +der(22)	3
+der(22), +8, +19	7
+der(22), +8,i(17q)	3
+der(22), +8,i(17q), +19	2

t(11;19)(q23;p13)	Biphenotypic acute leukemia
t(9;11)(p22;q23)	Biphenotypic acute leukemia
t(11;17)(q23;p13)	Biphenotypic acute leukemia
t(11;14)(q23;q32)	
del(12)(p11p13)	Common ALL L1/L2
Trisomy 6	
Trisomy 8	
Trisomy 18	
Trisomy 21	
Monosomy 20	
Hyperdiploidy >50 chromosomes	Early B precursor ALL; favorable prognosis, 30% children, 5% adults
Near haploid <30 chromosomes	Common ALL; very poor prognosis
Severe hypodiploidy 30–39 chromosomes	Mainly in adults

Acute lymphoblastic leukemia (ALL) incidence in childhood, 3 in 100 000 per year, peak age 3–5 years; in adults less than 30 years.

Table 9. FAB classification of ALL [3]

L1	Blasts are mainly small and relatively uniform in appearance. The nucleocytoplasmic ratio is high. Nuclei are predominantly round, nucleoli inconspicuous. Chromatin pattern is diffuse, smaller blasts show some chromatin condensation. Cytoplasm is scanty and slightly to moderately basophilic. Some cytoplasmic vacuolation may be present.
L2	Blasts are larger than L1, and more heterogeneous. Nucleocytoplasmic ratio is lower. Nuclei are more pleomorphic, with some nuclei being indented, cleft or irregular. Cytoplasm varies in amount but is often abundant. Cytoplasmic basophilia is variable. Cytoplasmic vacuolation may be present.
L3	Blasts are large and homogeneous. Nucleocytoplasmic ratio is high, though not as high as in L1. Nuclei are predominantly round with finely stippled chromatin pattern, and prominent nucleoli. Cytoplasm is strongly basophilic and, in at least some cells, there is heavy vacuolation.

The MIC classification for ALL exists in addition to FAB as for AML [7].

Table 10. Chromosome abnormalities in lymphomas

Abnormality	Association in lymphoma
Chromosome 1	25% of NHL often as secondary change
1p	T-cell lymphoma
1q	Diffuse large-cell lymphoma
t(2;5)(p23;q35)	Malignant histiocytosis
Trisomy 3	T-cell/diffuse mixed large- and small-cell lymphoma
Chromosome 3	25% of NHL/diffuse large-cell lymphoma
del(3p)	Immunoblastic lymphoma
6p	T-cell lymphoma
del(6q)	15% NHL often as secondary change
i(6p)	Follicular small, cleaved-cell-type lymphoma
Trisomy 7	5–15% NHL/ follicular large-cell lymphoma
7p15–21 7q35-36	T-cell lymphoma breakpoints are at sites where T-cell receptors have been mapped
Trisomy 8	Follicular, mixed-small-cleaved-cell and large-cell lymphomas
del(11)(q23–25)	B-cell immunophenotype, Diffuse mixed small- and large-cell lymphoma
t(11;14)(q13;q32)	Small-cell lymphocytic lymphoma, and B-cell CLL
Trisomy 12	Small-cell lymphocytic lymphoma, and B-cell CLL
t(14;18)(q32;q21)	Most common change in NHL. B-cell lymphomas of follicular morphology, frequent in small, cleaved-cell type
14q+	Most common in NHL (includes the t(14;18) and t(11;14) as described above) occurs in 50% of cases
14q11	T-cell lymphoma
17q21–25	Follicular large-cell lymphoma
Trisomy 18	10–15% NHL usually as secondary change
del(22)(q11)	Occurring as Philadelphia translocation variable histological type

CLL, chronic lymphocytic leukemia; NHL, non-Hodgkin's lymphoma.

28 *Human Cytogenetics*

Table 11. Chromosome changes in chronic lymphoproliferative disorders

Diagnosis	Abnormality
B lineage	
Chronic lymphocytic leukemia (CLL)	+12, 14q+, del(13q)
Prolymphocytic leukemia (PLL)	14q+, t/del(12)(p12–p13)
Hairy-cell leukemia (HCL)	14q+, del(14q)
Waldenström's macroglobulinemia (WM)	?
Multiple myeloma (MM)	Rearrangements chromosome 1, 14q+
Plasma-cell leukemia (PCL)	Rearrangements chromosome 1, 14q+
T lineage	
Large granular lymphocytic leukemia (LGLL)	inv(14)(q11q32), t/del(14)(q11)
Adult T-cell leukemia/lymphoma (ATL)	14q+, 14q11, del(6q)
Prolymphocytic leukemia	14q11
Cutaneous T-cell lymphoma, Sézary's syndrome, mycosis fungoides	Rearrangements chromosome 1, t/del(6p)

The classification of CLL and NHL is still based mainly on cytological, histological and immunophenotypic characteristics, although cytogenetics is becoming increasingly important. Certain cytogenetic abnormalities are associated with B- and T-cell disease, as can be seen in the above tables. There are at least six well-described histopathological systems used in varying institutions for classification [8–13].

Table 12. Translocations and oncogenes

Abnormality	Respective genes	
Acute myeloid leukemia (AML/ANLL)		
inv(3)(q21q26)		evi1
t(3;3)(q21;q26)		evi1
t(6;9)(p23;q34)	can	dek
t(8;21)(q22;q22)	eto	aml1
t(9;11)(p22;q23)		Trithorax
t(11;19)(q23;p13)	mll (Trithorax)	
t(15;17)(q22;q11)	rarα	pml
Acute lymphoblastic leukemia (ALL)		
Pre-B cell and B cell		
t(1;19)(q23;p13)	pbxi	e2α
t(2;8)(p12;q24)	Igκ	c-*myc*
t(8;14)(q24;q32)	c-*myc*	Ig heavy
t(8;22)(q24;q11)	c-*myc*	Igγ
t(17;19)(q22;p13)	hlf	e2α
Mixed		
t(4;11)(q21;q23)		mll Trithorax
t(9;22)(q34;q11)	abl	bcr
t(11;19)(q23;p13)	mll Trithorax	e2α

Continued

Table 12. Translocations and oncogenes, *continued*

Abnormality	Respective genes	
T cell		
t(1;7)(p34;q34)	tcrβ	lck
t(1;14)(p32;q11)	tcl5	tcrẟ
t(2;8)(q24;q24)	tcl4	c-*myc*
t(7;7)(p15;q11)	tcrγ	
t(7;9)(q34-36;q34)	tcrβ	tcl4
t(7;19)(q34-36;p23)	tcrβ	lyl1
t(8;14)(q24;q11)	c-*myc*	tcrα
t(10;14)(q24;q11)	tcl3	tcrẟ
t(11;14)(p13;q11)	tcl2	tcrẟ
t(11;14)(p15;q11)	tcl1	tcrẟ
inv(14)(q11q32)	tcrα	Ig heavy
Chronic lymphocytic leukemia (CLL)		
B cell		
t(2;14)(p13;q32)		Ig heavy
t(14;19)(q32;q13)	Ig heavy	bcl3
T cell		
t(8;14)(q24;q11)	c-*myc*	tcrα
inv(14)(q11q32)	tcrα	Ig heavy
Multiple myeloma		
t(11;14)(q13;q32)	bcl1	Ig heavy

Chronic myeloid leukemia (CML/CGL)		
t(9;22)(q34;q11)	abl	bcr
Non-Hodgkin's lymphoma		
t(2;8)(p12;q24)	Igκ	c-*myc*
t(8;14)(q24;q11)	c-*myc*	Ig heavy
t(8;22)(q24;q11)	c-*myc*	Igγ
t(11;14)(q13;q32)	bcl1	Ig heavy
t(14;18)(q32;q21)	Ig heavy	bcl2

ANLL, acute nonlymphocytic leukemia; CGL, chronic granulocytic leukemia.

Table 13. Common chromosome changes in solid tumors

Tumor	Abnormality	Incidence (%)
Lipoma	Changes 6p, 12q13–15, 13q, t(3;12)(q27-28;q14-15)	50
Atypical lipoma	Ring chromosome	30
Leioma of uterus	Changes 6p, 7q, 12q14–15 t(12;14)(q14-15;q23-24) del(7)(q21,q31), +12	?
Myxoid liposarcoma	t(12;16)(q13;p11)	?
Synovial sarcoma	t(X;18)(p11;q11)	?
Rhabdomyosarcoma	t(2;13)(q35-37;q14), +2	?
Dermatofibrosarcoma protuberans	Ring chromosome	?

Continued

Human Cytogenetics

Table 13. Common chromosome changes in solid tumors, *continued*

Tumor	Abnormality	Incidence (%)
Malignant fibrous histiocytoma	Changes 1, 3p, 11p	?
	t(19;?)(p13;?), ring	
Infantile fibrosarcoma	Trisomies	?
Extraskeletal myxoid chondrosarcoma	t(9;22)(q22;q12)	?
Ewing's sarcoma/Askin tumor/peripheral neuroepithelioma	t(11;22)(q24;q12)	> 90
Salivary gland: adenoma	Changes 3p21, 8q12, 12q13–15	?
	t(3;8)(p21;q12)	
Adenocarcinoma	del(6q)	?
Squamous cell carcinoma of head and neck	Changes 1p22, 11q13	?
Lung carcinoma	del(3p)	?
Large bowel carcinoma	Changes 1	20
	Trisomy 7	30
	Trisomy 12	10
	Changes 17, 18	
Renal carcinoma	t or del(3)(p11–21)	80
	t(5;14)(q13;q22)	?
Wilms tumor	Changes 1	50
	t or del(11)(p13)	30
Bladder carcinoma	Changes 1	30
	i(5p)	20
	Trisomy 7	10
	Monsomy 9	10

	Changes 11	30
	3p	
Breast carcinoma	Changes of 1	80
	t or del(16q)	20
	Changes of 3, 11 and 17	
Ovarian carcinoma	Changes of 1	80
	t(6;14)(q21;q24)	
	11p, t(19;?)(p13;?)	
Benign ovarian tumors	Trisomy 12	
Germ-cell tumors of testis	i(12p)	90
Prostatic adenocarcinoma	del(7)(q22)	?
	del(10)(q24)	?
Meningioma	Monosomy 22, del(22q)	>90
Malignant glioma	dmin	50
	Changes of 1p, 9p, 17p, +7, −10, −22, −X, −Y	
Neuroblastoma	del(1)(p13–32)	70
	HSR/dmin	70
Retinoblastoma	Changes of 1	50
	i(6p)	30
	del(13)(q14)/−13	20
Malignant melanoma	t or del(1)(p12–22)	60
	t(1;19)(q12;p13)	?
	t or del(6q)/i(6p)	80
	+7	50
Uterine carcinoma	Changes of 1	80

HSR, homogeneously staining regions; dmin, double minutes.

34   *Human Cytogenetics*

Table 14. Symptoms of diseases encountered

Condition and hematology findings	Symptoms
Acute leukemia	
Normochromic, normocytic anemia	Due to marrow failure:
WCC decreased, normal or increased	Pallor, lethargy, anemia
Thrombocytopenia (extreme in AML)	Fever, malaise, features of infections, including septicemia
Variable numbers of blast cells in blood film. AML films may contain Auer rods, and other abnormal cells may be present, promyelocytes, agranular neutrophils, myelomonocytic cells	Spontaneous bruises, purpura, bleeding gums and bleeding from venepuncture sites due to thrombocytopenia
Hypercellular bone marrow, marked proliferation of blast cells, typically over 75% of the marrow cell total	Due to organ infiltration:
Disseminated intravascular coagulation in AML M3	Tender bones, especially in children
	Superficial lymphadenopathy in ALL
	Moderate splenomegaly, hepatomegaly (ALL)
	Gum hypertrophy and infiltration, rectal ulceration, skin involvement (particularly AML M4 and M5 types)
	Meningeal syndrome (ALL), headache, nausea and vomiting
	Testicular swelling (ALL)
	Mediastinal compression (particularly T-cell ALL or T-lymphoblastic lymphoma)
Chronic granulocytic leukemia (CGL) and chronic myeloid leukemia (CML)	
Leukocytosis usually $> 50 \times 10^9 \ l^{-1}$ up to $500 \times 10^9 \ l^{-1}$	Hypermetabolism, e.g. weight loss, lassitude, anorexia and night sweats
Complete spectrum of myeloid cells in PB. The levels of neutrophils and myelocytes exceed the blast cells and promyelocytes	Splenomegaly nearly always present and sometimes massive. The enlargement can cause discomfort, pain or indigestion
Hypercellular marrow with granulopoietic predominance	

Increased circulating basophils Platelet count normal, decreased or increased Neutrophil alkaline phosphatase (NAP) score low	Pallor, dyspnea and tachycardia, features of anemia Bruising, epistaxis, menorrhagia or hemorrhage from other sites

Chronic lymphocytic leukemia (CLL)
  Leukocytosis between 5 and 300×10^9 l^{-1}. 70–90% of white cells on blood film appear as small lymphocytes
  Normocytic, normochromic anemia
  Thrombocytopenia
  BM shows lymphocytes comprising 25–95% of all cells
  Reduced concentration of serum immunoglobulins
	Symmetrical enlargement of superficial lymph nodes
	Pallor, dyspnea
	Splenomegaly and hepatomegaly
	Bruising in patients with thrombocytopenia
	Pruritus associated with herpes zoster
	Tonsillar enlargement

Hairy-cell leukemia (HCL)
  The 'hairy cells' (a type of B lymphocyte) are present in blood, marrow, liver and other organs
  Trephine shows mild fibrosis
  Serum paraprotein may be present
	Spleen moderately enlarged
	Pancytopenia
	Disease peak at 40–60 years of age, with a male to female ratio of 4:1

Myelodysplastic syndromes (MDS) Wide range of abnormalities in PB and BM: macrocytosis, ring sideroblasts, megaloblastic erythropoiesis, disordered granulopoiesis and megakaryocytes	Qualitative and quantitative abnormalities in one or more of the three myeloid cell lines: red cells, granulocytes and monocytes and platelets Anemia Infections due to impaired phagocytic production and/or function (See *Table 3* for further breakdown of this group)

Continued

Human Cytogenetics

Table 14. Symptoms of diseases encountered, *continued*

Condition and hematology findings	Symptoms
Hodgkin's disease (HD)	
Normochromic, normocytic anemia, with marrow failure and infiltration	Painless, nontender asymmetrical, firm, discrete enlargement of superficial lymph nodes
Leukocytosis in one-third of patients due to neutrophil increase	Splenomegaly in 50% patients. The liver may be enlarged
NAP score increased	Mediastinal involvement in 6–11% patients (nodular sclerosis type in women)
Eosinophilia	
Lymphopenia (advanced disease)	Cutaneous HD occurs as late complication
Platelet count normal or increased in early disease but low in later stages	Also seen: fever, pruritis, alcohol-induced pain, weight loss, profuse night sweats, weakness and fatigue
ESR raised	
BM involvement rare in early disease	
Non-Hodgkin's lymphoma (NHL)	
Normochromic, normocytic anemia, also autoimmune hemolytic anemia may develop	Median presentation age 50 years
	Superficial lymphadenopathy
When BM involved, neutropenia, thrombocytopenia or leukoerythroblastic features	Fever, night sweats and weight loss less frequent than in HD and usually indicate disseminated disease. Anemia and infections
Lymphoma cells may be present in PB	
Trephine shows focal involvement in about 20% of cases. Diffuse infiltration and fibrosis may occur	Oropharyngeal involvement, sore throat, obstructed breathing in 5–10% patients
	Abdominal disease, liver and spleen often enlarged

Burkitt's lymphoma (BL)
    Isolated histiocytes in masses of abnormal lymphocytes produce the 'starry sky' appearance in tissue sections
    Epstein–Barr virus identified in Burkitt cell culture

Mycosis fungoides and Sézary's syndrome
    Circulating T-lymphocyte cells
    Cutaneous T-cell lymphoma

Multiple myeloma
    98% patients monoclonal protein occurs in serum and/or urine
    Bence-Jones protein in two-thirds of cases
    BM shows increased plama cells
    Normochromic, normocytic or macrocytic anemia
    Rouleaux formation
    Neutropenia and thrombocytopenia in advanced cases

Continued

Skin, brain and testis or thyroid involvement. The skin is also primarily involved in two closely related T-cell lymphomas, mycosis fungoides and Sézary's syndrome

B-cell lymphoblastic lymphoma
Predominantly in young African children
Massive jaw lesions
Extranodal abdominal involvement
Ovarian tumors (in girls)

Severe pruritis and psoriaform lesions
Lymph nodes, spleen, liver and bone marrow ultimately affected
Exfoliative dermatitis
Erythroderma
Generalized lymphadenopathy

Bone pain (especially backache)
Lethargy, weakness, dyspnea, pallor, tachycardia due to anemia
Repeated infections caused by deficient antibody production and later due to neutropenia
Anorexia, vomiting, constipation and mental disturbance due to renal failure

Table 14. Symptoms of diseases encountered, *continued*

Condition and hematology findings	Symptoms
PB film shows abnormal plasma cells (15% patients) Serum calcium elevation (45%) Blood urea raised (20%) Low serum albumin in advanced disease	Abnormal bleeding tendency: myeloma proteins interfere with platelet function and coagulation factors
Waldenström's macroglobulinemia	
Proliferation of cells which produce monoclonal IgM paraprotein Blood viscosity increased High ESR PB lymphocytosis BM infiltration by small lymphocytes, plasma cells, 'plasmacytoid' forms, immature lymphoid cells, mast cells and histiocytes	Seen mostly in males over 50 years of age Fatigue and weight loss Hyperviscosity syndrome Engorged veins in retina Bleeding tendency Anemia due to hemodilution, decreased red cell survival, blood loss, BM failure Moderate lymphadenopathy, enlargement of liver and spleen
Polycythemia rubra vera	
Hemoglobin, hematocrit and red cell count increased Neutrophil leukocytosis (in over half of patients) Raised platelet count (in half of patients) NAP score increased Increased serum vitamin B_{12} binding capacity	Headaches, pruritis, dyspnea, blurred vision and night sweats Retinal venous engorgement, conjunctival suffusion Splenomegaly (in two-thirds of patients) Hemorrhage or thrombosis Gout

Hypercellular bone marrow with prominent
    megakaryocytes
  Blood viscosity increased

Essential thrombocythemia
  Abnormal large platelets and megakaryocyte fragments in
    PB film
  Platelet count raised above 1000
  Platelet function tests abnormal

Anemia
Massive splenomegaly giving discomfort, pain or indigestion
Weight loss, anorexia and night sweats
Bleeding problems and bone pain

ALL, acute lymphoblastic leukemia; AML, acute myeloid leukemia; BM, bone marrow; ESR, erythrocyte sedimentation rate; PB, peripheral blood; WCC, white cell count.

Chapter 3 THE RELATIONSHIP BETWEEN CYTOGENETIC AND MOLECULAR GENETIC STUDIES – J. Waters and F. MacDonald

Human genetic disorders have traditionally been divided into two subgroups: chromosomal disorders and single gene disorders. This subdivision reflected the apparent gap that existed between gross chromosomal abnormalities visible by light microscopy on the one hand and mutations amenable to molecular analysis using a variety of techniques (e.g. polymerase chain reaction (PCR), Southern blotting) on the other. For a number of reasons this subdivision has broken down:

1. The ability to screen for and characterize 'chromosomal disorders' (e.g. Prader–Willi syndrome, Fragile-X syndrome) by molecular means;
2. The advent of new or improved techniques, most notably fluorescence *in situ* hybridization (FISH), which allow

Table 1 lists genetic disorders where such complementary tests may be required. By its very nature such a list has its limitations:

1. It is not feasible to list all mendelian disorders occasionally associated with chromosome deletion or rearrangement. This area has recently been covered by Tommerup [2]. The reader should also refer to ref. 3.
2. An exhaustive list of nonrandom chromosomal rearrangements in leukemias and solid tumors has not been included. Once the relevant genes have been cloned all of these are amenable to a complementary molecular approach involving the use of FISH, PCR or Southern blotting to analyze DNA rearrangements directly.

submicroscopic rearrangements (deletions, duplications) to be visualized on metaphase chromosomes and, equally importantly, in interphase cells; and

3. A greater understanding that a large number of 'mendelian' single gene disorders may arise as a result of partial monosomy ('haploinsufficiency syndromes') [1].

The clinical cytogeneticist needs to be increasingly alert to the possibility of arranging complementary tests, either initiated by himself/herself (e.g. FISH) or complementary molecular tests that will require liaison with molecular geneticists using the same sample.

Finally, it is worth emphasizing that the time is rapidly approaching when the entire human genome will be mapped with YAC (yeast artificial chromosomes) contigs. Using a FISH-based approach, all cytogenetic abnormalities will be accessible to further characterization at the molecular level using a battery of probes, both YACs and cosmid subclones. Furthermore, comparative genomic hybridization (CGH) offers the prospect of screening the entire chromosome complement for partial monosomy/trisomy, however generated, offering the exciting prospect of new findings in molecular cytogenetics.

Human Cytogenetics

Table 1. Cytogenetic and molecular genetic approaches to disease

Chromosome	Band	Disease	Chromosome abnormality	Applicability of FISH	Probe ID// Candidate gene	Molecular test	Reference
1	p34	T-cell ALL (acute lymphocytic leukemia)	t(1;14)(p34;q11) (commonest rearrangement)	Confirmation of breakpoints/ interphase screening for minimal residual disease	//TAL1 (#1);TCRd (#14)	RT-PCR approach described	4
1	q23	pre-B-cell ALL (acute lymphocytic leukemia)	t(1;19)(q23;p13.3) (commonest rearrangement, see t(17;19))	Confirmation of breakpoints/ interphase screening for minimal residual disease	//PBX1 (#1);E2A (#19)	RT-PCR approach described	5
1	q32–q41	Van der Woude syndrome 1	Microdeletions described	?			6
2	p21	Holoprosence-phaly 2	Microdeletions described	?			7
2	q34–q36	Van der Woude syndrome 2	Microdeletions described	?			1
3	p25–pter	3p25-pter deletion syndrome	Terminal deletion	Confirmation by FISH possible		Detection of deletions and parental origin using RFLP and microsatellite markers on 3p by comparison with parental alleles	8
3	q22–q23	Blepharophimosis– Ptosis–Epicanthus inversus	Microdeletions described	?			2

Cytogenetic and Molecular Genetic Studies

3	q24–q25	Dandy–Walker syndrome	Microdeletions described	?			1
3	p25	Familial renal cell carcinoma	t(3;8)(p21;q24)	Possible, to define breakpoints		Candidate gene HRCA-1	9
4	p16.3	Wolf–Hirschhorn syndrome	Large terminal deletions and translocations in majority, submicroscopic deletions in minority	Yes	Many probes for breakpoints FISH, e.g. D4S95, D4S96, D4S97	Detection of deletions using microsatellite and RFLP markers possible by comparison with parental alleles; candidate gene, ZNF141, recently isolated	10,11
4	q12-13	Piebald trait	Microdeletions described	?			12
4	q21	Infant ALL (acute lymphocytic leukemia)	t(4;11)(q21;q23)(commonest rearrangement)	Confirmation of breakpoints/ interphase screening for minimal residual disease	//FEL (AF-4) (#4); HRX (MLL) or ALL1 (#11)	RT-PCR detection of HRX-FEL fusion transcripts described	13
5	p15	Cri-du-chat syndrome	Large terminal deletions and translocations in majority, submicroscopic deletions in minority	Yes	D5S23 cosmid//		14
5	q21	Familial adeno-matous polyposis (FAP)	1) Deletions and rearrangements in up to 10% of families. 2) Remainder of families have point mutations	Not usually applicable	Many probes and microsatellites around the APC gene locus	1) Southern blot and PCR analysis with specific gene probes or linked markers to detect deletion or rearrangement 2) Direct mutation detection	15–17
6	p23	AML (acute myeloid leukemia)	t(6;9)(p23;q34)	Confirmation of breakpoints/ interphase screening	//DEK(#6),CAN(#9)	RT-PCR approach described	18

Continued

Table 1. Cytogenetic and molecular genetic approaches to disease, *continued*

Chromosome	Band	Disease	Chromosome abnormality	Applicability of FISH	Probe ID// Candidate gene	Molecular test	Reference
7	p21-p22	Saethre–Chotzen syndrome	53% of cases have deletion of 7p; few cases of translocations also reported	Yes		Linkage analysis with markers on 7p	19
7	p13	Greig cephalopoly-syndactyly syndrome (GCPS)	Translocations and deletions in up to 50% of cases	?	//GL1-3	Linkage analysis possible using RFLPs in EGFR; gene now identified as the GLI-3 gene; mutation analysis may become possible	20–23
7	p11.2–p14	Craniosynostosis syndrome 1	Microdeletions described	?			2,24
7	q11.2	Williams syndrome	Microdeletions described	Yes	cELN-272 and cELN-11D//Elastin	RFLPs and micro-satellite markers in the elastin gene to identify allele loss in affected individuals if parental samples available; dosage to identify hemizygosity possible on Southern blot analysis	25,26

7	q11.23	Zellweger syndrome	Microdeletions in proximal long arm breakpoint in two cases of inv(7)	?			27
7	q36–qter	Holoprosencephaly 3	Microdeletions described	?		RFLPs can be used to map deletion	28
8	p11.1	Spherocytosis type II (ankyrin defect) syndrome	Microdeletions described within ankyrin gene	?	//Ankyrin		29
8	q11–q13	Branchio-oto-renal syndrome	Microdeletions described	?			A.O. Wilkie, pers.comm.
8	q22	AML-M2 (acute myeloid leukemia)	t(8;21)(q22;q22.3)	Confirmation of breakpoints/ interphase screening for minimal residual disease	//ETO(#8),AML1(#21)	RT-PCR approach	30
8	q24.11	Trichorhinophalangeal syndrome (TRP1)	Deletions of q23.3–q24.13; inversions (TRP1)	?			2,31
8	q24.11	Langer–Giedion syndrome (TRP2); Trichorhinophalangeal syndrome (TRP1) maps to same region	Deletions of q24.11–q24.13 described	?		RFLPs from 8q24 can be used to characterize deletion/parental origin if parents samples available; possible by dosage if index case only available; presence of heterozygosity or particular marker indicates no deletion	32,33

Continued

Human Cytogenetics

Table 1. Cytogenetic and molecular genetic approaches to disease, *continued*

Chromosome	Band	Disease	Chromosome abnormality	Applicability of FISH	Probe ID// Candidate gene	Molecular test	Reference
9	q21.2–q23	Mental retardation with growth delay	Microdeletions described	?			1
9	q22	Fanconi's anemia		Not applicable		Direct mutation analysis of FACC gene becoming possible; linkage analysis	34
9	q33	Goltz-like syndrome	Microdeletions described	?			1
9	q34	CML (chronic myeloid leukemia)	t(9;22)(q34;q11) (commonest rearrangement)	Confirmation of breakpoints/ interphase screening for minimal residual disease	//ABL(#9),BCR(#22)	RT-PCR approach described. BCR rearrangements detectable by Southern blot analysis	35,36
10	p13	DiGeorge syndrome 2 (DGCR2)	Microdeletions described	?			37
10	q11.2	Hirschsprung's disease	Microdeletions described	?	//RET	Mutation analysis of RET proto-oncogene possible	1
11	p15.5	Beckwith– Wiedemann syndrome	Duplication of distal part of p15 (paternal origin); *de novo* balanced translocations (maternal origin); viable deletions not described	?		Linkage analysis possible in familial cases	2

11	p13	Wilms tumor (WT); WAGR (WT, aniridia, genital anomalies, mental retardation)	Deletion of whole or part of p13 (some submicroscopic); *de novo* balanced translocations; 2–10% have chromosomal rearrangements; visible deletions	Yes	//WT1 (Wilms tumor susceptibility gene); AN2/PAX6 (aniridia 2); loci are 700–1000 kb apart	Linkage analysis in familial cases; detection of deletions by dosage or PFGE; direct detection of mutations in WT1	2,38–42
13	q12.2	Moebius syndrome		?			43
13	q14	Retinoblastoma	Deletion of q14 (cytogenetically visible in 3–5% of cases); also translocations involving this band and point mutations in RB1 gene	Yes	Submicroscopic deletions at 13q14 detectable by FISH: Rb-cosmid #//RB1	Linkage analysis in families; cDNA probes to screen for submicroscopic deletions by Southern or PFGE analysis; direct mutation detection in remainder of cases	44–47,
14	q32	Follicular lymphoma	t(14;18)(q32;q21) (commonest rearrangement)	Confirmation of breakpoints /interphase screening for minimal residual disease	//IgHL(#14), BCL-2(#18)	RT-PCR approach described	F. Cottar, pers. comm.

Continued

Human Cytogenetics

Table 1. Cytogenetic and molecular genetic approaches to disease, *continued*

Chromosome	Band	Disease	Chromosome abnormality	Applicability of FISH	Probe ID// Candidate gene	Molecular test	Reference
15	q12	Prader–Willi syndrome (PWS)/ Angelman syndrome (AS)	60–70% show microdeletion at 15q12	Confirmation of deletion	SNRPN* (proximal PWS/AS critical region)GABRB3* (distal PWS/AS critical region)	Screening for methylation status within critical region combined with CA repeat linkage analysis allows characterization of deletions and uniparental disomy	48–51
15	q22	AML-M3 (acute promyelocytic leukemia)	t(15;17)(q22;q11)	Confirmation of breakpoints/ interphase screening for minimal residual disease	//MLL(#16), RAR(#17)	RT-PCR approach described	52
16	p13.3	α-Thalassemia/ mental retardation syndrome (ATR-16)	Deletions of 16p13.3 (1–2 Mb) have been described; cases with no deletions may have X-linked form (ATR-X)	Possible		RFLPs from 16p13.3 can be used to characterize deletion if parents samples available; possible by dosage if index case only available; PFGE can be used to characterize deletion	53

16	p13	Rubinstein–Taybi syndrome (RSTS)	Submicroscopic deletion in 25% of cases	Yes		RFLPs from 16p13 used in conjunction with parental samples to map deletion/parental origin	54
16	p13	AML-M4Eo, M2 (acute myeloid leukemia)	inv(16)(p13q22), del(16q), other rearrangements involving 16p13 and 16q22	Can use combination of 16het and distal 16p cosmids to confirm inversion	YAC 55.3 (spans 16p13 breakpoint) //Myosin heavy chain (#16p13)transcription factor#(16q22)	RT-PCR approach possible	55
17	p13.3	Miller–Dieker syndrome (MDS)	90% of cases have deletion involving LIS-1 gene; 10% of cases with isolated lissencephaly also have deletion	FISH will detect both MDS and isolated lissencephaly	Cosmid D17S379#// LIS-1	RFLPs from 17p13 used in conjunction with parental samples to map deletion in both MDS and isolated cases of lissencephaly; cDNA for LIS-1 gene available to detect deletions	56–59
17	p11.2	Charcot–Marie–Tooth 1A syndrome	1) Duplication in PMP-22 gene in 65–85% 2) Remainder have point mutation	FISHable using combined metaphase/ interphase screening	CMT1A-cosmid*// PMP-22	1) Duplication detected by dosage/presence of three alleles 2) Direct mutation analysis	54,60,61

Continued

Human Cytogenetics

Table 1. Cytogenetic and molecular genetic approaches to disease, *continued*

Chromosome	Band	Disease	Chromosome abnormality	Applicability of FISH	Probe ID// Candidate gene	Molecular test	Reference
17	p11.2	Smith–Magenis syndrome	Interstitial deletion of 17p11.2	Yes	D17S29#//	Detection of deletions using microsatellite and RFLP markers on 17p possible by comparison with parental alleles; dosage analysis possible if index case only available	55
17	q11	Neurofibroma-tosis 1	1) Few cases of translocations involving q11 2) Majority have mutations in NF1 gene	Translocation cases may be associated with submicroscopic deletions (< 30 kb) detectable by FISH	//NF1	1) cDNAs for NF1 to detect rearrangements by Southern/PFGE analysis 2) Direct mutation analysis available for point mutations; linkage available for families	62–65
17	q22	ALL (acute lymphocytic leukemia)	t(17:19)(q22;pB.3) Rare (approx. 1% of ALL cases)	Confirmation of breakpoints/ interphase screening for minimal residual disease	//HLF (#17);E2A (#19)	RT-PCR approach described	66

17	q24–q25	Campomelic dysplasia (CMPD1) 46,XY sex reversal (SRA1) (?contiguous deletion syndrome)	Few cases of translocations, inversions described; microdeletions not yet described in CMPD1/SRA1	?			2
20	p11.23–p12.1	Alagille syndrome	Interstitial deletions of p arm in some cases			Linkage analysis possible	67–69
21	q22.3	Down syndrome	Trisomy	FISHable	Centromeric alphoid (13/21)# Contigs/YACs available for Down syndrome critical region (DSCR) at 21q22.2–q22.3#	Many microsatellite repeats available on chr. 21 to determine parent of origin of extra copy; quantitative analysis using microsatellite markers can be used to diagnose presence of trisomy directly	70–72
22	q11.21–q11.23	DiGeorge syndrome 1/ CATCH 22	Cytogenetically visible deletion in 5–10% of cases	FISHable DiGeorge critical region (DGCR) is 2 Mb in size; most deletions span this region; translocations with 22q22.2 breakpoints have been described	D22S75#, DO832, Sc4.1, Sc11.1// Tuple-1	Allele loss by RFLP analysis if parental DNA available; dosage on Southern blot analysis if only affected individual available	73–77

Continued

Table 1. Cytogenetic and molecular genetic approaches to disease, *continued*

Chromosome	Band	Disease	Chromosome abnormality	Applicability of FISH	Probe ID// Candidate gene	Molecular test	Reference
X	p22.3	Steroid sulfatase deficiency	Intragenic deletions	Yes		Linkage analysis; cDNA analysis by Southern blotting for deletions	78
X	p22.3	Kallman syndrome	Intragenic deletions and point mutations	Yes		cDNA analysis by Southern blotting for large deletions; PCR/direct sequencing for point mutations	79,80
X	p21	Chronic granulomatous disease (CYBB)	Intragenic deletions	?			81
X	p21	Duchenne muscular dystrophy (DMD)	Some deletions cytogenetically detectable; submicroscopic deletions in 50–60% of affected boys; duplications, point mutations in remainder; few cases of affected females show translocations involving DMD gene	Yes for carrier females		cDNAs to characterize rearrangements by Southern blotting or PFGE; other smaller rearrangements detected by PCR analysis; RFLP analysis and RT-PCR used for carrier detection	37,82–85

Chromosome	Band	Disease/Gene	Abnormality	Application	Probe order for FISH	Additional notes	Ref
X	q27.3	Fragile-X syndrome (FRAXA)	Fragile site (FRAXA) at Xq27.3 (trinucleotide expansion) seen in a percentage of cells in most affected males and half obligate carrier females; few cases of point mutations or deletions of FMR-1 gene	To distinguish from FRAXE	Probe order for FISH: G9L-FRAXA-141R-FRAXE-VK21//FMR-1	Detection of expansion of trinucleotides possible by Southern blotting with probes StB12.3 or Ox1.9 and Ox0.55; PCR useful for excluding disease and for identification of premutation	86,87
X	q27.3	Fragile X with mild mental retardation (FRAXE)	Fragile site (FRAXE) at Xq27.3 distal to FRAXA	To distinguish from FRAXA		PCR analysis; multiplex FRAXA/FRAXE possible	88
X	q27.3	Fragile X (FRAXF)	Fragile site (FRAXF) at Xq27.3/q28 distal to FRAXE	To distinguish from FRAXA/FRAXE			89
Y	p11.3	SRY	X–Y interchanges, dicentric Y; Y-derived markers	Characterization of XX males, XX true hermaphrodites, XY females	//SRY		90,91
Y	p11.3	ZFY	X–Y interchanges, dicentric Y; Y-derived markers	Characterization of XX males, XX true hermaphrodites, XY females	pMF-1//ZFY		91

EGFR, epidermal growth factor receptor; PFGE, pulsed field gel electrophoresis; RFLP, restriction fragment length polymorphism; RT, reverse transcription.

Chapter 4 ANALYZING CHROMOSOMES: STAINING, BANDING AND MICROSCOPY – A.J. Monk

The microscope is the chief investigative tool of the cytogeneticist. A basic requirement of microscopy is to give contrast and resolution with suitable magnification. This can be achieved using a number of staining methods (see *Table 1*) in combination with two main microscopy techniques. These are transmitted light (bright field/phase contrast microscopy) and fluorescence microscopy. A high quality research-standard microscope adjusted correctly is essential to achieve optimum resolution of chromosomes. Information is given here to aid in the understanding of the types of lenses available, as well as a protocol for microscope adjustments (see *Figure 1*). A brief explanation of the principles of reflected light fluorescence microscopy is also included, along with information on the fluorochromes commonly used in cytogenetics (*Table 2*).

Obtaining a permanent record of one's observations by photography is largely a matter for personal investigation, but some basic recommendations are included in *Table 3*.

Computerized imaging systems are adding another dimension to the work of the cytogeneticist. This is a competitive field, developing very rapidly, with a number of systems on the market offering a range of facilities. Recently the emphasis has been on imaging systems for fluorescence *in situ* hybridization (FISH) work, and a section is included outlining some of the problems encountered.

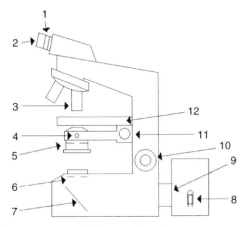

Figure 1. Arrangement of light source, lens elements and controls on the bright field microscope. 1, Eyepiece focusing ring (diopter adjustment); 2, eyepiece; 3, objective lens; 4, substage condenser centering screw; 5, aperture iris diaphragm; 6, field iris diaphragm; 7, mirror; 8, halogen light source; 9, collector lens; 10, fine and coarse focusing controls; 11, substage condenser height control; 12, specimen stage.

1 Microscope adjustment for bright field observation

1.1 Eyepiece adjustment (interpupillary distance)

1. Place a slide on the microscope stage and bring the image into focus using a low power objective;
2. Adjust the interpupillary distance of the eyepieces until the left and right eye images merge into one.

Microscopes from different manufacturers may differ in that either one or both eyepieces may be capable of independent focusing.

One adjustable eyepiece

1. Look at the image through the nonadjustable eyepiece, blanking off the other with a piece of card. Bring the image into sharp focus using the fine focusing control on the microscope stand.
2. Look at the image through the adjustable eyepiece (blanking off the other) and use the diopter focusing

adjustment on this eyepiece, to bring the image into sharp focus.

Two adjustable eyepieces

There will be an engraved line or '0' mark on the diopter adjustment ring. Both eyepieces should be set to this position before proceeding as described for 'One adjustable eyepiece'.

1.2 Condenser adjustment (for Köhler illumination)

For correct Köhler illumination, the image of the field iris diaphragm should be in the same focusing plane as the image from the specimen. This is achieved as follows:

1. Using the field iris diaphragm control, reduce the diameter of the field diaphragm to its smallest size.
2. Use the condenser height control to bring an image of the field diaphragm into focus.
3. Bring the image of the field diaphragm into the center of the field of view using the two condenser centering screws.

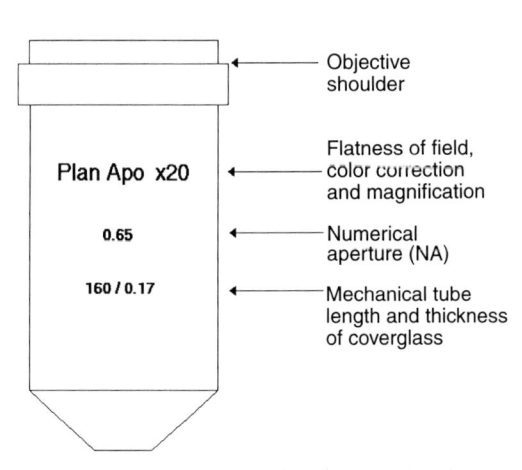

Figure 2. Common engravings found on an objective lens.

4. Widen the diameter of the field iris diaphragm progressively until the polygonal image just fits inside the field of view. The condenser is now centered.
5. The field iris diaphragm may now be opened slightly to take its image just out of the field of view.

1.3 Aperture iris diaphragm adjustment

In order to achieve optimum objective resolution the aperture iris diaphragm should be set to the numerical aperture of the objective in use. This number is engraved on the objective lens body (*Figure 2*). However, stopping down the aperture diaphragm slightly more than indicated by the numerical aperture value will result in better image contrast and increased depth of field.

If at this point we remove one of the eyepieces and look directly into the observation tube, an image of the aperture iris diaphragm will be seen, giving approximately 80% of the full field.

2 Microscope adjustment for phase contrast observation

Phase contrast microscopy requires the use of a condenser which has built-in annular stops (light rings). The size of the annular stop is matched to the size of a dark ring found inside a corresponding phase contrast objective lens. For each objective a different annular stop is selected by rotating a plate on the condenser. One position of this rotating plate usually allows bright field illumination.

1. Select the bright field position on the condenser and proceed as for Section 2.2.
2. With a phase contrast objective in position, select the appropriate annular stop on the condenser rotating plate.
3. Open the aperture iris diaphragm fully.
4. Place a specimen on the microscope stage and focus.
5. Remove one of the eyepieces and insert a centering telescope. Focus its eyepiece until the light and dark phase rings are visible (*Figure 3a*).

6. Insert suitable adjusting keys into the annular stop plate and turn them until the light and dark rings coincide (*Figure 3b*).
7. Repeat this process for the other objectives lens and annular stop combination.
8. Remove the centering telescope and replace the eyepiece.

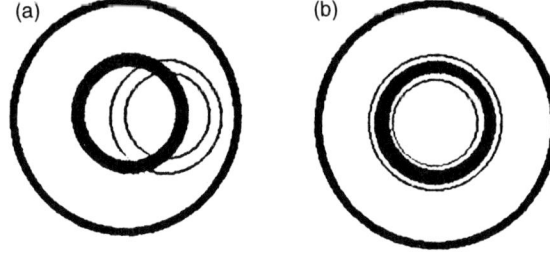

Figure 3. Arrangement of annular rings observed by phase contrast microscopy. (a) Before correct centering of the phase plate; (b) correctly centered.

achromats. It is corrected in the same way, but the component wavelengths have been brought closer together. These lenses are so called because they contain elements made from the mineral 'fluorite'.

Apochromats (Apo)
These are the most highly corrected lenses, color errors have been, for all practical purposes, eliminated. Lenses of this type are ideal for color photography.

3.3 Magnification
This indicates the objective magnification not the total magnification of the optical system. The eyepieces provide a further stage of magnification to the final image.

3.4 Numerical aperture
This value is a measure of the light-gathering ability of a lens. The more light a lens can gather, the brighter and sharper the image becomes. In turn, this leads to an increase in the

3 Objective lens markings

3.1 Flatness of field
Plan
Lenses of this type give a flat field of view without blurring at the periphery of the visual field.

3.2 Color correction
Achromatic
These are the simplest type of lenses, designed to place red and blue light into a common focus, with green light brought into the shortest focus. Green is the brightest part of the spectrum and the region in which the human eye has greatest discriminatory power.

Lenses of this type may give slight red/blue color fringes around the edges of a specimen so a green filter should be used. If there is no mention of color correction on the lens, it can be assumed to be of this type.

Fluorite or semi-apochromats (FL, Fluor)
Color error in these lenses is more highly corrected than with resolving power of a lens. The higher the numerical aperture, the higher the resolving power for a given magnification. Numerical apertures of greater than 1.0 require the use of oil, or some other immersion medium. Lenses which require immersion are also marked Oil or OEL or HI (homogeneous immersion).

3.5 Mechanical tube length
160 (finite optics)
The distance from the shoulder of the objective (see *Figure 2*) to the top of the sleeve containing the eyepiece, in millimeters, is known as the mechanical tube length.

To make it possible to use a number of objectives and eyepieces with different magnification ratios a number of physical dimensions of the microscope have become standardized. The mechanical tube length is generally 160 mm.

∞ (infinite optics)
Until recently the beams of light passing through the

objective were made to converge, forming an intermediate image inside the microscope tube which was then further magnified by the eyepiece for observation (*Figure 4a*).

The modern research microscope often requires the addition of intermediate attachments between the objective and the eyepiece (e.g. fluorescence illuminators and filters).

This often requires the use of additional corrective lenses and leads to unwanted magnification due to changes in the mechanical tube length. In turn this may lead to a reduction in brightness and overall resolution of the image.

In an infinity optical system the light from the specimen travels from the objective in parallel beams (*Figure 4b*). The intermediate image is formed where the beams are made to converge when passing through a 'Telan' lens. Intermediate attachments can be added in the region of parallel light beams without corrective lenses and unwanted magnification.

3.6 Thickness of coverglass
The performance of high magnification, large numerical aperture dry objectives can vary significantly depending on whether or not a coverglass is used.

0.17
Indicates an objective to be used with a 0.17 mm thick coverglass.

0, Epi, NCG
Indicates an objective suitable for uncovered specimens.

-
Objectives marked in this way are not affected by the presence or absence of a coverglass.

Some objectives may be fitted with a correction collar to allow for small variations in the thickness of a coverglass.

3.7 Other markings
Ph, P, PC
Objectives marked in this way are designed for phase

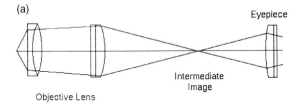

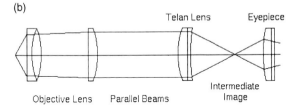

Figure 4. Arrangement of lens elements and light paths in finite optics (a) and infinite optics (b).

contrast microscopy.

U, UV

This indicates that the lens is designed to be used with UV illumination, e.g. fluorescence microscopy.

4 The fluorescence microscope

Fluorochromes are dyes which absorb light of one particular wavelength, then re-emit light of a longer wavelength. Different fluorochromes are distinguished by their absorption and emission spectra. *Table 2* lists the absorption and emission spectra for a number of fluorochromes commonly used in cytogenetics.

A fluorescence microscope uses this phenomenon, by means of an arrangement of filters, to reveal certain features within the specimen. The most usual arrangement of light source and filters is shown in *Figure 5*.

A powerful light source, usually a mercury vapor lamp, is required to fully 'excite' the fluorochrome, and accurate adjustment of this light source is essential. The method of adjusting the light source may vary between manufacturers so the microscope documentation should be consulted for this information.

Three filters are required for the microscope to function, usually arranged in a single unit. Each of these filters has a specific task to perform (*Figure 6*).

1. Excitation filters select the optimum wavelength for the fluorochrome to absorb.
2. Barrier filters suppress the excess excitation light and select out the emitted wavelengths of the fluorochrome.
3. The dichroic mirror or chromatic beam splitter, reflects light shorter than a certain wavelength and transmits light of a longer wavelength.

This arrangement of light source and filters has a number of advantages. Firstly, as only one lens acts as both condenser and objective, alignment between the two is unnecessary.

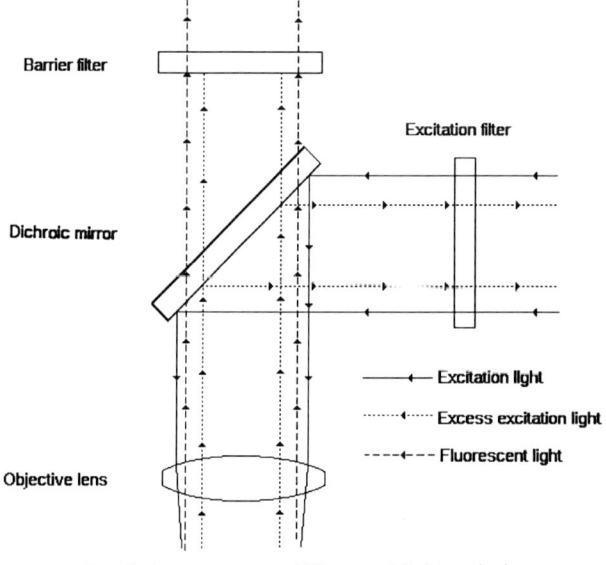

Figure 6. Detailed arrangement of filters and light paths in a reflected light fluorescence microscope.

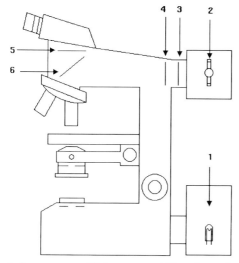

Figure 5. Arrangement of light source and filters in a reflected light (epi-) fluorescence microscope. 1, Bright field light source; 2, mercury vapor lamp; 3, heat absorption filter; 4, excitation filter; 5, barrier filter; 6, dichroic mirror.

Secondly, as the substage illuminator is undisturbed, it is easy to change over to bright field microscopy.

5 Computerized imaging systems

The application of computerized imaging systems in cytogenetics has undergone considerable development in recent years and is still changing rapidly.

These systems represent the next stage on from the electronic darkroom (see Section 7). Once the image has been captured by a computer-controlled camera it can be subjected to a number of image manipulation or enhancement techniques, including karyotyping. Some of these machines even offer the facility to scan slides and find metaphases.

In recent years a number of these machines have been specifically tailored for FISH imaging. The FISH technique often requires more than one fluorochrome to be visualized simultaneously on the same specimen and this has revealed a limitation of the fluorescence microscope.

Physically changing from one filter unit to another can cause a shift in the image position known as registration error. The amount of registration error varies from one microscope to another, with the Olympus BX system showing apparently very little. A number of solutions to this problem have been developed, including 'triple band pass' filter blocks which allow up to three fluorochromes to be visualized simultaneously.

Triple filter blocks are not an ideal solution, as they can reduce the intensity of the final image, a bright counterstain can obliterate small signals of a different color and some mixing of colors can occur due to overlap of the absorption/emission spectra for the fluorochromes involved.

Other solutions have involved using computers to capture separate images for each fluorochrome then combining these to give a full color, error-free image.

Registration error correction is usually achieved in one of two ways:

(metaphase finding/karyotyping, FISH imaging and laboratory management system), Perceptive Scientific International Ltd (metaphase finding/karyotyping, FISH imaging and laboratory management), Imagenetics (karyotyping, FISH imaging and laboratory database system), Oncor (Alpha Laboratories Ltd; FISH imaging and laboratory database system), Optivision (FISH imaging and image archiving database), Nikon Instruments (UK) Ltd (FISH imaging and database) and Leica Cambridge Ltd (karyotyping and FISH imaging).

6   Photography

6.1   Conventional 35 mm photography

In general there is a direct relationship between a film's sensitivity (A.S.A. rating) and the quality of the final recorded image. Where a high quality photomicrograph is important and long exposure times can be tolerated, a low A.S.A-rated film will give better results.

1. Using a triple band pass dichroic mirror and barrier filter but discrete excitation filters mounted on a rotating wheel. This avoids changing the parts of the filter system which are placed in the light path between the objective and the eyepiece; or
2. Using discrete filter blocks for each fluorochrome and then using a software offset to realign the images into one.

Due to the number of configurations available on these systems, prices can vary considerably. Careful examination of your requirements is recommended.

A number of companies are currently offering machines capable of metaphase finding, karyotyping and/or FISH imaging. These include: Applied Imaging International Ltd

Fluorescence photomicroscopy can present problems as the light level of the microscope image is low and the fluorochromes tend to fade rapidly. A compromise between exposure time and final image quality has to be made (see *Table 3*).

6.2 Electronic darkroom

Methods of capturing images using 'TV' cameras then reproducing a 'hard copy' with a video printer have developed considerably in recent years.

The prints produced are of photographic quality, do not fade and are very quick to produce. Capital outlay for a system varies depending upon the application (see *Table 3*).

Table 1. Staining methods

Technique	Staining pattern	Applications	Mechanism	Limitations and comments
Solid stain	Uniform dark appearance	Useful for studying chromosome breakage, or dicentrics and fragile sites	—	Largely obsolete due to banding techniques. Produces sharply defined chromosomes with no significant swelling
Q-banding	Series of bright and dull fluorescent bands. The distal region of the Y chromosome is particularly bright	Useful for routine chromosome identification, and for studying polymorphisms associated with chromosome 3 and 4, the acrocentrics and the Y	AT-rich DNA enhances the fluorescence of Quinacrine. CG-rich DNA tends to quench this fluorescence	Requires the use of a fluorescence microscope with UV illuminator The fluorescent pattern tends to fade during observation so photography is essential
G-banding	Light and dark stained regions along the length of the chromosome. Grossly similar to Q-bands	Most widely used technique for routine staining of mammalian chromosomes	The mechanism of G-banding is not fully understood. AT-rich DNA appears darkly stained, CG	The bands obtained by this method are of higher resolution than fluorochrome methods and are permanent. High resolution culture and

			lightly stained. The darkly stained bands correlate with the chromomeres seen at pachytene in meiosis, generally replicate their DNA late in the S-phase and appear to contain relatively few active genes	banding techniques can produce in excess of 1500 bands per cell. Results can be variable depending upon a number of factors, e.g. tissue type, slide making, slide aging, pH of buffers, etc.
R-banding	Light and dark stained regions along the length of the chromosome. Generally the reverse pattern to that obtained with either Q- or C-banding	Several clinical laboratories prefer R-bands to G-bands for routine cytogenetic investigation	A number of methods exist for producing an R-banding pattern, involving the treatment of slides at high temperatures in various buffers then staining with fluorochromes or Giemsa. The chemical basis for this staining reaction is unclear	One advantage to this staining procedure is that the telomeric regions of several chromosomes are darkly stained. Disadvantages to this technique include the use of fluorescence microscopy or the problems encountered with G-banding

Continued

Table 1. Staining methods, *continued*

Technique	Staining pattern	Applications	Mechanism	Limitations and comments
C-banding	Dark staining of constitutive heterochromatin located at the centromere of all chromosomes and the distal portion of the long arm of Y. Chromosomes 1, 9 and 16 have larger C-bands than do other chromosomes	The C-bands of chromosomes may vary significantly between homologs and individuals. Polymorphisms of this type are familial and can be used as markers. C-bands are also useful in studying chromosomes of unusual morphology, or marker chrromosomes	In C-banding, the chromosomal DNA is preferentially denatured in alkali and lost from the non-C-banding regions. The reasons for this are not clear but may involve histono proteins closely bound to the C-band heterochromatin	Slides may require aging before C-banding. The timing of the denaturation is critical, and may vary between tissue types
Restriction endonuclease/ Giemsa banding	Treatment of slides with restriction enzymes prior to staining with Giemsa generally gives a modified C-banding pattern depending upon the choice of enzyme. A banding pattern similar to G-banding can be obtained with *HaeIII*	Some small polymorphisms cannot be delineated by C-banding, particularly those of chromosomes 4 and 18. However, this technique is able to highlight even minor polymorphisms. The characteristic patterns produced by different enzymes may be useful	Endonuclease enzymes have short recognition sequences which cleave DNA at specific sites. The chromosomal regions which have a large number of sites for a particular enzyme will lose DNA preferentially from these regions. The loss of DNA is	These techniques are easy to use, and can be applied to fresh slides, allowing observation in a few hours. Preparations are permanent and produce little distortion of chromosome morphology

			directly related to a decrease in Giemsa staining	
Silver staining (NOR staining)	Silver staining appears as one or more dot-like structures of varying sizes. These are found on the 'stalks' of the acrocentric chromosomes (13, 14, 15, 21, 22) not the satellites	Applications of silver staining are numerous. The pattern obtained is consistent within an individual and is heritable. Silver staining is a crucial tool in discriminating small bisatellited marker chromosomes from other markers of similar appearance	The nucleolar organizer regions (NOR) are known to contain the genes for 18S and 28S fractions of ribosomal RNA. Silver impregnation is thought to identify a protein adjacent to the NOR not the NOR itself	This technique can sometimes be a little unreliable, with some batches of chemicals working better than others. One advantage of this technique is that other staining procedures can be carried out after silver staining
G-11 banding	Chromosomes stain blue but the centromeric regions of chromosomes 1, 3, 5, 7, 9, 10, 19 and the Y chromosome stain magenta. Chromosome 9 shows a particularly strong reaction	The technique has been used to demonstrate polymorphisms and pericentric inversions of chromosome 9. It is also used to identify chromosomes in man–mouse somatic cell hybrids, as the rodent chromosomes stain uniformly magenta	The selectivity of Giemsa staining at pH 11 is not clearly understood	At pH 11 the Giemsa staining solution tends to precipitate out, which may cause deposition of eosin on the slide. However the rate of this precipitation slows after about 10 min. Cytoplasm on the slides may also interfere with the results, as it tends to remain blue

Continued

Table 1. Staining methods, *continued*

Technique	Staining pattern	Applications	Mechanism	Limitations and comments
Distamycin A–DAPI banding (DA–DAPI)	Staining with DA-DAPI reveals bright fluorescence in the heterochromatic regions of chromosomes 1, 9, 16, the distal long arm of the Y, and the proximal short arm of chromosome 15	This technique is particularly useful in identifying small satellited markers derived from chromosome 15	Both DA and DAPI have an affinity for A–T base pairs, binding at similar but not identical sites. Competition between the two ligands may lead to differential fluorescence	The fluorescent banding pattern tends to fade during observations, making photography essential
T-banding	This technique produces fluorescent bands at the most distal terminal bands of the chromosome arms	Useful for studying rearrangements involving the telomeres of chromosomes, which may not be visible under conventional G-banding	T-banding can be considered as a special case of R-banding where more aggressive denaturation treatments leave staining only at the terminal bands	—

Kinetochore staining	A number of staining techniques identify pairs of dots at the centromeres of chromosomes. These may represent the kinetochore itself or chromatin associated with it	This technique has been used for investigating kinetochore inactivation in dicentric chromosomes	Kinetochore staining can be achieved in a number of ways including fixation and aging regimes, or immunofluorescence using antibodies obtained from scleroderma patients	—
X chromatin identification	This technique reveals a darkly stained body usually at the periphery of interphase nuclei. This is known as a Barr body	Useful as a noninvasive method for rapid screening of sex chromosome constitution without elaborate chromosomal analysis	The Barr body represents the inactive X chromosome, which remains condensed throughout interphase in female cells. The number of Barr bodies in any one cell is equal to the number of X chromosomes minus 1	Buccal smears are the most readily available source of material for sex chromatin analysis. They are easy to take and the staining procedure is quick. However, 100 or 200 cells should be examined as sex chromatin is not seen in every female cell, usually in only 15–30%

Continued

72   *Human Cytogenetics*

Table 1. Staining methods, *continued*

Technique	Staining pattern	Applications	Mechanism	Limitations and comments
Y chromatin identification	A brightly fluorescent body in the nucleus of interphase cells	—	—	When buccal smears are stained with quinacrine dihydrochloride, the brightly fluorescent spot found in the nucleus corresponds to the bright band in the distal long arm of the Y chromosome seen in metaphase preparations (see Q-banding, p. 66)
Differential replication staining	The staining pattern obtained depends upon the timing of introduction of 5-bromodeoxyuridine (BrdU) in relation to the cell cycle. Late introduction of BrdU gives a pattern similar to R-banding. Early introduction gives a pattern similar to G-banding	This method has been used to investigate how different parts of mammalian chromosomes replicate at different times in the cell cycle	BrdU is a thymidine analog which is readily incorporated into chromosomes during DNA replication. If the BrdU-substituted chromosomes are then exposed to UV light the BrdU breaks down. This in turn affects the way in which Giemsa binds to the chromosomes, resulting in differential staining	Cultures must be incubated in the dark after the addition of BrdU. If BrdU is present for two full cell cycles, one chromatid will have BrdU substituted into one polynucleotide chain, the other will have BrdU substituted in both chains. This gives the harlequin staining used for detecting sister chromatid exchanges

Table 2. Fluorochromes commonly used in cytogenetics[a]

Fluorochrome	Excitation range (nm)	Emission range (nm)
Coumarin	340–380 (UV)	430–500
DAPI	320–380 (UV)	480–520
Quinacrine mustard	420–440 (violet/blue)	500–510
Acridine Orange	470 (blue)	530–650
FITC	490–500 (blue)	510–530
Rhodamine	530–560 (green)	580–625
Propidium iodide	530–560 (green)	590–610
Texas Red	568 (green)	610–620

[a] For bright field use a monochromatic green filter (e.g. Wratten 58 (Kodak)) or a 550 nm interference filter.

Table 3. Photographic equipment and materials

Recommended equipment/materials	Comments
Films for conventional 35 mm photomicroscopy Kodak Technical Pan 2415 25 A.S.A. rated black and white negative film developed using HC-110	This is a particularly flexible film for general light microscopy as different levels of contrast can be obtained by variations in the development regime
Films for fluorescence photomicroscopy Kodak Ektachrome Elite 400 A.S.A. color slide film	Slide films rather than print films are generally used for fluorescence photography as they seem to give better contrast with 'black' backgrounds
Kodak Ektachrome 160 Tungsten Kodak Ektachrome 320 Tungsten These two films are slide films color-balanced for tungsten light rather than daylight	For publications it is possible to obtain prints directly from slides without any intermediate stages using R3 processing (Chromagene Ltd; see Chapter 9 for address)
Cameras Sony XC-77CE (monochrome) Sony DXC-930P (color integrating camera) Hamamatsu C5310 series (chilled color camera) Both of the color cameras mentioned here are suitable for fluorescence microscopy	These systems have a number of significant advantages over conventional photography, i.e. 1. They require no darkroom facilities; 2. They give an almost instantaneous result; 3. Computerized image archiving facilities can be added at a later date
Video printer Sony UP5200 (A5 print size) Sony UP7000 (A4 print size) Mitsubishi CP2000E (A4 print size) Hitachi VY-300E (A5 print size)	

Acknowledgments

AJM would like to thank Olympus Optical Co. (UK) Ltd for their technical help and Mrs B. Herod for typing the manuscript.

Chapter 5 **CULTURING CELLS FOR CHROMOSOME PREPARATION –**
D.E. Rooney and B.H. Czepulkowski

Cell and tissue culture methods for chromosome preparations are many and varied, and merit much more detail than we have space available for here. Besides which, textbooks of methodology already exist [1,2]. We have, therefore, endeavored to concentrate on what we may regard as 'useful reminders' which either prompt the memory as to which reagent concentration/culture regime to use, or act as a source of revision for the trainee.

We start with a summary of the principal media available to us as cytogeneticists (*Table 1*) alongside a reminder of the various concentrations of reagents commonly used for routine purposes (*Table 2*). We have not extended this list to the more specialized applications, such as the expression

Tables 5–8 relate to the culture of bone marrow and leukemic blood, perhaps the greatest challenge to the cytogeneticist in terms of the actual growing of the sample. Here, more than anywhere, the culture regime and correct seeding density of the cells is crucial; expression of a malignant clone may be dependent upon the proper application of technique. *Table 5* shows the normal hematological values against which a count supplied on a request form may be compared, while *Table 6* provides at-a-glance seeding volumes for given white cell counts.

Bone marrow is the tissue of choice for cytogenetic study for most hematological disorders, although blood can be useful in cases where large numbers of nucleated cells are present, of chromosome breakage syndromes, or *in situ* hybridization as these topics require a more extensive approach. However, a revision list of the conditions of expression for the chromosome instability syndromes (and a few other atypical culture conditions) is included (*Table 3*) at the risk of overlapping slightly with *Table 1* in Chapter 7.

As far as prenatal diagnosis is concerned, cell culture is pretty straightforward, the interpretation being more controversial, so we have confined ourselves to a brief summary of the principal types of cells obtained from cultured amniotic fluid (*Table 4*) as summarized by Gosden [3].

such as chronic myeloid leukemia (CML), chronic lymphocytic leukemia (CLL) and high white cell count acute lymphoblastic leukemia (ALL). *Table 7* shows the recommended cultures for the various leukemic and pre-leukemic conditions, with an indication of the likelihood that white cells may be present in large numbers. Remember that overseeding of nucleated cells in bone marrow culture is far more detrimental than under-seeding! Finally, *Table 8* lists some of the referral reasons likely to be encountered on request forms for samples where diagnosis has not yet been reached. If the information in this table is cross-referred to that in *Table 7*, the appropriate culture regime may be applied even if a final diagnosis has not yet been obtained.

Table 1. Media commonly used for cell and tissue culture for cytogenetic study

Medium	Recommended uses	Other information
Medium 199 (TC 199)	L	Depleted folate levels. May be used for AF and F although not normally the medium of choice
Minimal essential medium (Eagle's)	L	A basic medium containing HBSS or EBSS according to buffering system required. Supplementation with nonessential amino acids and vitamins required
Dulbecco's modified MEM	L	As above, but already supplemented with amino acids and vitamins
Iscove's	L	A modification of Dulbecco's MEM designed for lymphocyte culture without serum which is replaced by bovine serum albumin, human transferrin and soybean lipid. Folate-free formulation for expression of fragile sites
FX-1	L	Folate-free medium for the expression of fragile sites
Ham's nutrient mixture F10	(L), CV, AF, F,	Good general-purpose medium. Particularly favored for long-term cultures
RPMI 1640	L, BM, (AF, CV, F)	Good general-purpose medium, Dutch modification particularly favored for lymphocyte and bone marrow cultures, for which it was originally formulated. Low thymidine content makes it a suitable medium for methotrexate, FdU and thymidine block synchronization cultures
Liebovitz L-15	F	Designed for growth of human fibroblasts in the presence of serum, but without CO_2
McCoy's 5A	BM	General-purpose medium which is favored by some laboratories for bone marrow cultures
Chang medium	AF, CV	Originally formulated for rapid growth of amniotic fluid cells, this medium has found wider application in the long-term culture of chorionic villus mesenchyme cells

AF, amniotic fluid culture; BM, bone marrow culture; CV, chorionic villus culture; F, fibroblast culture; L, lymphocyte culture; MEM, minimal essential medium.

Reproduced from ref. 1 with permission from Churchill Livingstone.

Table 2. Reagents commonly used for routine cell and tissue culture for cytogenetic study

Reagent	Uses	Stock solution	Working solution	Final concentration
General				
L-Glutamine	Nutrient, unstable at 37°C	200 mM	200 mM	1%
Sodium bicarbonate	Buffer	7.5%	7.5%	1%
Hepes buffer	Buffer	1 M	1 M	1%
Versene (EDTA)	Dispersal of cell monolayer	1:5000	1:5000	1–2 ml neat
Trypsin	Dispersal of cell monolayer	0.25% (1:250)	0.25% (1:250)	0.5–1ml neat
Colcemid	Mitotic arrest	10 µg ml^{-1}	10 µg ml^{-1}	0.5–1%
Heparin	Anticoagulant for blood and bone marrow specimens	5000 IU ml^{-1}	5000 IU ml^{-1}	0.3 ml neat
Dimethylsulfoxide (DMSO)	Cell freezing	Neat	Neat	14%
Glycerol	Cell freezing	Neat	Neat	10%
Antibiotics and fungicides				
Nystatin	Fungicide particularly effective against *Candida* spp.	10 000 IU ml^{-1}	10 000 IU ml^{-1}	0.5%
Penicillin	Broad-spectrum antibiotic	5000 or 10 000 IU ml^{-1}	5000 or 10 000 IU ml^{-1}	1 or 2%
Streptomycin	Broad-spectrum antibiotic	5000 or 10 000 IU ml^{-1}	5000 or 10 000 IU ml^{-1}	1 or 2%
Mitogens				
Phytohemagglutinin (PHA)	T-cell mitogen	10 ml reconstituted, lyophilized	2 µg ml^{-1}	1%
PWM	B- and T-cell mitogen	5 ml reconstituted, lyophilized	1 µg ml^{-1}	1%

Continued

Table 2. Reagents commonly used for routine cell and tissue culture for cytogenetic study, *continued*

Reagent	Uses	Stock solution	Working solution	Final concentration
TPA (PMA)	B-cell mitogen	100 µg ml^{-1} in DMSO or absolute alcohol	2–5 µg ml^{-1}	50 ng ml^{-1}
Synchronizing agents				
MTX	Cell synchronization	1 mg/2.2 ml	1 µg ml^{-1}	10^{-7} M (0.1 ml per 5 ml culture)
FdU	Cell synchronization	10 µg ml^{-1}	0.1 ml/10 ml distilled water	4×10^{-3} M (0.1 ml per 5 ml culture)
Thymidine	(1) For cell synchronization	15 mg ml^{-1}	15 mg ml^{-1}	2%
	(2) for release of FdU block	2.5 mg ml^{-1}		2%
Uridine	Required for release of FdU block	10 mg ml^{-1}	20 ml in 10 ml distilled water	3×10^{-3} M (0.1 ml per 5 ml culture)

EDTA, ethylenediaminetetraacetic acid.

Table 3. Culture conditions for expression of specific chromosome abnormalities and methods of analysis

Abnormality	Conditions of expression	Methods of analysis
Chromosome instability syndromes		
Ataxia telangiectasia (AT)	Induction of chromatid aberrations by G_2 X-irradiation. Exposure to radiomimetic chemicals, e.g. bleomycin, streptonigrin, tallysomycin	Increase in chromatid-type damage. May be 10- to 20-fold differential between AT and normal controls following irradiation
Bloom's syndrome	BrdU incorporation into cultured cells over two complete cycles Exposure to ethylating agents, e.g. ethyl methane sulfonate	Increased sister chromatid exchange and spontaneous aberrations
Fanconi's anemia	Exposure to alkylating agents, e.g. mitomycin C, dipoxybutane, nitrogen mustard	Induced multiple aberrations
ICF syndrome	Standard lymphocyte culture	Uncondensed constitutive heterochromatin in the paracentromeric regions of chromosomes 1, 9 and 16. Seen as stretching, breaking and association of centromeres of these chromosomes, and multibranched chromosomes, particularly of chromosome 1
Nijmegen breakage syndrome	Exposure to X-rays and bleomycin	Induced aberrations. Spontaneous abnormalities of chromosome 7 often involving chromosome 14
Robert's syndrome	Standard lymphocyte culture + C-banding	Premature centromeric division, manifesting as 'puffing' of paracentromeric regions of chromosomes 1, 9, 16 and Yq. Centromeric C-bands appear as paired dots

Continued

Table 3. Culture conditions for expression of specific chromosome abnormalities and methods of analysis, *continued*

Abnormality	Conditions of expression	Methods of analysis
Xeroderma pigmentosum	Exposure to UV irradiation	Induced aberrations
Other syndromes		
Fragile X	Low folate levels, folate antagonists and thymidylate stress	Fragile site at Xq(27), always appears as mosaic. Mosaic cell counts indicated
Pallister–Killian syndrome	Preferential expression in fibroblast cultures	Isochromsome 12(q) either absent or at very low levels of mosaicism in lymphocyte cultures. Mosaic counts in fibroblast cultures indicated
Di George syndrome	Use B-cell mitogen or set up lymphoblastoid cell line if T-cell response poor	Lack of T lymphocytes may impair quality of preparations. High-resolution banding required to resolve this microdeletion of 22q11

BrdU, bromodeoxyuridine; ICF, immunodeficiency, centrometric heterochromatin instability and facial anomalies.

Table 4. Principal types of cultured amniotic fluid cells and their properties

	Epithelioid (E) cells	Amniotic fluid (AF) cells	Fibroblastic (F) cells
Morphology			
Derivation	Squames/epithelial fetal skin, bladder and other epithelia	Intermediate fetal membrane and trophoblast	Fibroblast-like fibrous connective tissue and dermal fibroblasts
Proportion	20%	>70%	<10%
Growth characteristics	Small colonies/sheets of highly coherent cells	Loose meshwork of loosely associated cells	Parallel arrays of cells
Cloning efficiency	Poor	Intermediate	Excellent
Mitoses	Rare	Moderately frequent	Frequent
Mean cell doubling	Lowest	Intermediate	High
Trypsinization	Resistant	Intermediate	Repeated subculture
Multinucleates	Some	Intermediate	Rare
Cell characteristics			
hCG production	−ve	++ve	−ve
Collagen production	−ve	Type IV ++	Types I and III ++
Collagen fibers	None	None	Network type I fibers
Basement membrane glycoprotein	None	Epithelial basement membrane components	—
Fibronectin	−ve	++	+
Desmosome complexes	Formed in regions of cell to cell contact	None	None
Keratin	+ve	+ve	−ve staining
Vimentin	−ve (epithelial origin)	+ve (30% of cells epithelial; some mesenchymal)	++ve staining (cells mesenchymal)

hCG, human chorionic gondatrophin.
Data reproduced from ref. 1 with permission from Churchill Livingstone.

Table 5. Normal hematological values

	Units	Males	Females
Hemoglobin	g dl^{-1}	13.5–17.5	11.5–15.5
Red cells (erythrocytes)	$\times 10^{12}$ l^{-1}	4.5–6.5	3.9–5.6
PCV (hematocrit)	%	40–52	36–48
Red cell mass		30 + 5 ml kg^{-1}	25 + 5 ml kg^{-1}
Plasma volume		45 + 5 ml kg^{-1}	45 + 5 ml kg^{-1}
MCV (mean corpuscular volume)	fl	80–95	
Platelets	$\times 10^9$ l^{-1}	150–400	
White cells normal blood counts	$\times 10^9$ l^{-1}		
Adults			
Total leukocytes (white cells)		4.00–11.0	
Neutrophils		2.50–7.5	
Eosinophils		0.04–0.4	
Monocytes		0.20–0.8	
Basophils		0.01–0.1	
Lymphocytes		1.50–3.5	
Children, total leukocytes			
Neonates		10.0–25.0	
1 year old		6.0–18.0	
4–7 years old		6.0–15.0	
8–12 years old		4.5–13.5	

Table 6. Bone marrow sample seeding volumes according to white cell count

WCC ($\times 10^6$ ml^{-1})	Volume cell suspension (ml)	Volume medium (ml)
1	Maximum	3–4
5	1	4
10	0.5	4.5
15	0.33	4.67
20	0.25	4.75
50	0.1	5
100	0.05	5
200	0.025	5
500	0.01	5
1000	0.005	5

Final cell concentration = 10^6 ml^{-1}.
Final volume = 5 ml.
Reproduced from ref. 2 with permission from Oxford University Press.

Human Cytogenetics

Table 7. Guidelines for setting up bone marrow samples

Diagnosis	Nucleated cells	Cultures
Acute lymphoblastic leukemia (ALL)	White cell count varies, CARE required!	O/N, O/N + colcemid, 24 h synchronized
Acute myeloid leukemia (AML)	Usually low, but care is required here	O/N, O/N + colcemid, 24 h synchronized
Acute promyelocytic leukemia (APML)	Low to medium	O/N, O/N + colcemid, 24 h synchronized (direct uninformative)
Myelodysplastic syndrome (MDS: RAEB, RA, CMML, RARS, RAEBT)	Low count except for CMML which is high	O/N, O/N + colcemid, 24 h synchronized, 2-day sometimes useful
Myeloproliferative disorders MPD (PRV, ET, MF)	Medium to high, care required	O/N, O/N + colcemid, 24 h synchronized
Chronic myeloid or granulocytic leukemia (CML, CGL)	Always high. Use less marrow than you think required	Direct, O/N, O/N + colcemid, 24 h synchronized
Lymphoproliferative disorders (unspecified)	Medium to high	O/N + colcemid, 3–5 day, 3–5 day + TPA or PWM
Chronic lymphocytic leukemia (CLL) B cell	High, care as for CML and CGL	O/N + colcemid, 3–5 day, 3–5 day + TPA or PWM
Prolymphocytic leukemia (PLL) B cell	Medium to high	O/N + colcemid, 3–5 day, 3–5 day + TPA or PWM
Hairy-cell leukemia (HCL) B cell	Very slow growing	O/N + colcemid (variable), 3–7 day, 3–7 day + TPA or PWM
Waldenström's macroglobulinemia (WM)	Medium	O/N + colcemid, 3–5 day, 3–5 day + TPA, or PWM
Plasma cell leukemia (PCL)	Medium	O/N + colcemid, 3–5 day, 3–5 day + TPA or PWM
Myeloma, or multiple myeloma	Medium	O/N + colcemid, 3–5 day, 3–5 day + TPA or PWM
Lymphoma (non-Hodgkin's lymphoma NHL)	Medium	O/N + colcemid, 3–5 day 3–5 day + TPA or PWM
Hodgkin's disease (HD)	Slow growing	O/N + colcemid, 3–7 day, 3–7 day + TPA or PWM
Large granular cell leukemia LGLL (T cell)	Medium	O/N + colcemid, 3–5 day, 3–5 day + PHA
Adult T-cell leukemia/lymphoma (ATL)	Medium to high	O/N + colcemid, 3–5 day, 3–5 day + PHA

Prolymphocytic leukemia (PLL) T cell	High	O/N + colcemid, 3–5 day, 3–5 day + PHA
Hairy-cell leukemia (HCL) T cell	Medium (slow growth)	O/N + colcemid, 3–7 day, 3–7 day + PHA
Cutaneous T-cell lymphoma/Sézary's syndrome/mycosis fungoides	Medium	O/N + colcemid, 3–5 day, 3–5 day + PHA

O/N, overnight culture.
Full protocols in refs 2 and 3.

Table 8. Reasons for referral of bone marrow samples with potential diagnoses

Indications	Potential diagnoses
Blasts in blood (> 50% myeloblasts or lymphoblasts)	Acute leukemia (AML or ALL)
Disseminated intravascular coagulation (DIC)	AML M3
Eosinophilia (increase in eosinophils)	Eosinophilic leukemia (rare), or in association with other acute leukemias, HD
Neutrophil leukocytosis (increase in neutrophils)	Neoplasia of all types, lymphoma, melanoma, MPD, CGL and PRV
Neutropenia (decrease in neutrophils or can be associated with pancytopenia)	Bone marrow failure due to malignant disease, myeloma
Pancytopenia (decrease in red cells, granulocytes and platelets)	Aplastic anemia, leukemia, MDS, myeloma
Leukoerythroblastic reaction (erythroblasts as well as primitive white cells in blood)	Myeloma, lymphoma, AML
Monocytosis (increase in monocytes)	HD, myelomonocytic leukemia, CMML
MCV raised	Megaloblastic anemia
Lymphocytosis (increase in lymphocytes)	CLL, lymphoma

Continued

Table 8. Reasons for referral of bone marrow samples with potential diagnoses, *continued*

Indications	Potential diagnoses
Lymphopenia (decrease in lymphocytes)	Rare, HD, bone marrow failure
Thrombocytopenia (decrease in platelets)	Extreme in AML. CLL, MDS, myeloma
Leukocytosis (increase in total white cells)	AML, ALL, CML, CGL, CLL, MPD
Reduced serum immunoglobulins	CLL
Anemia	MDS
ESR raised	HD, myeloma, WM
Presence of serum paraprotein	Myeloma, CLL, lymphoma, WM
Thrombocytosis (increase in platelets)	MPD, ET
Splenic enlargement	CGL, CML, ET, myelofibrosis

Chapter 6 PRENATAL DIAGNOSIS OF CHROMOSOME ABNORMALITY –

D.E. Rooney

Prenatal diagnosis of virtually all microscopically visible chromosome abnormalities is possible, but by far the most important application is the detection of Down syndrome, due to the association with maternal age. For many years, the only method of prenatal diagnosis of chromosome disorders was by means of amniocentesis at 16 weeks of pregnancy or more. The past decade has seen the development of several other approaches, some of which have dramatically increased the detection of Down syndrome and the other major trisomies. *Table 1* gives estimates of rates per 1000 of chromosome abnormalities in live births by single-year interval of maternal age.

1 Screening methods

1.1 Maternal age
Until recently, maternal age, normally 35 years and above, has been the principal indication for a prenatal chromosome test. However, since the majority of births occur in the lower age group, most cases of Down syndrome and the other major trisomies will not be detected prenatally using age as the sole criterion. *Tables 2–4* summarize data correlating risks and incidences of the major trisomies with maternal age.

1.2 Maternal serum biochemistry
Three biochemical markers of Down syndrome, when correlated with maternal age and weight, significantly increase detection, namely:

1. Raised maternal serum human chorionic gonadotrophin (hCG);
2. Low maternal serum unconjugated estriol (uE$_3$);
3. Low maternal serum alphafetoprotein (AFP).

These data, together with gestational age and several other lesser factors, provide a risk figure, which is computer-generated. A prenatal test may be offered depending on the local risk cut-off. *Table 5* shows the detection rate (DR) and false positive rate (FPR) for a range of maternal ages, and risk cutoffs. DR versus FPR is dependent on the accuracy of the estimation of the gestational age. *Table 6* shows the variation in DR and FPR, as well as the odds of being affected given a positive screening result (OAPR), with weight and methods of estimating gestational age. At present, maternal serum screening is restricted to the second trimester of pregnancy.

1.3 Ultrasound

The ever-increasing sophistication of ultrasound technology provides another tool for fetal screening. There are presently two parameters which may be indicative of Down syndrome:

1. Nuchal thickening;
2. Decreased femur length.

Long-term CVS cell cultures [2,3] derive from the mesenchyme core, whereas the direct method (= semi-direct and short-term culture) represents the cytotrophoblast. Thus karyotypes obtained are cell-lineage-specific, and different chromosome complements may occasionally be obtained if both methods are applied simultaneously to the same sample. The abnormal cell line may or may not represent the fetal chromosome complement. This phenomenon plus confined placental mosaicism (when the placenta contains chromosomally abnormal cells not present in the fetus) [4,5] leads to equivocal results in approximately 2% of CVS diagnoses [6,7].

Mosaicism is also encountered in 1–2% of amniocenteses, though only 1–3 per 1000 represent true fetal mosaicism [8–10], the rest being pseudomosaicism. *Table 8a* and *b* shows the classification of patterns of mosaicism in CVS and amniotic fluid culture. *Table 9* summarizes the conclusions which have been drawn from several large CVS studies as to the potential significance and value of chromosome results from chorionic villi.

Table 7 summarizes some of the main chromosome abnormalities associated with other fetal anomalies detectable on ultrasound scan.

2 Methods of obtaining a fetal karyotype

2.1 Amniocentesis

Until recently, amniocentesis was carried out exclusively in the second trimester of pregnancy. Early amniocentesis (EA) is now available in some centers at 10 weeks' gestation, although most prefer a cutoff of 12 weeks.

2.2 Chorionic villus sampling and placental biopsy

The development of chorionic villus sampling (CVS) has enabled prenatal diagnosis to be performed in the first trimester of pregnancy. Although it can be offered as early as the 8th week of pregnancy, testing is normally offered at gestations of 10 weeks or above. Placental biopsy, and subsequent karyotyping by a direct method [1], enables a rapid result to be obtained from late second and third trimester pregnancies when anomalies have been detected by ultrasound scanning.

2.3 Fetal blood sampling

Fetal blood sampling can be performed from the 18th week of pregnancy and is normally indicated for karyotyping pregnancies with an abnormal scan result, or as a follow-up technique where prenatal diagnosis by another means has given an equivocal result.

Table 10 shows the percentage mosaicism excluded at three different confidence levels for cell counts commonly used. This is based on cells from a uniform suspension: bloods, bone marrows and trypsinized flask cultures from tissues and cells. Further information regarding cell counts from colonies obtained by *in situ* harvest methods may be found in refs 11 and 12.

3. Fetal measurement

From the 7th to the 14th week of pregnancy, crown–rump length (CRL) is the fetal dimension of preference for estimation of gestational age, although biparietal diameter (BPD), femur and humerus lengths are also measured after the 10th week. After 14 weeks' gestation, head circumference is used together with the other parameters (except CRL).

Tables 11 and *12* summarize these measurements (which may appear on cytogenetic request forms and fetal post-mortem reports).

Table 1. Estimates of rates per thousand of chromosome abnormalities in live births by single-year interval

Maternal age (years)[b]	Down syndrome	Edwards syndrome (trisomy 18)	Patau syndrome (trisomy 13)	XXY	XYY	Turner syndrome genotype	Other clinically significant abnormality[a]
< 15	1.0[c]	< 0.1[c]	< 0.1–0.1	0.4	0.5	< 0.1	0.2
15	1.0[o]	< 0.1[c]	< 0.1–0.1	0.4	0.5	< 0.1	0.2
16	0.9[c]	< 0.1[c]	< 0.1–0.1	0.4	0.5	< 0.1	0.2
17	0.8[c]	< 0.1[c]	< 0.1–0.1	0.4	0.5	< 0.1	0.2
18	0.7[c]	< 0.1[c]	< 0.1–0.1	0.4	0.5	< 0.1	0.2
19	0.6[c]	< 0.1[c]	< 0.1–0.1	0.4	0.5	< 0.1	0.2
20	0.5–0.7	< 0.1–0.1	< 0.1–0.1	0.4	0.5	< 0.1	0.2
21	0.5–0.7	< 0.1–0.1	< 0.1–0.1	0.4	0.5	< 0.1	0.2
22	0.6–0.8	< 0.1–0.1	< 0.1–0.1	0.4	0.5	< 0.1	0.2
23	0.6–0.8	< 0.1–0.1	< 0.1–0.1	0.4	0.5	< 0.1	0.2
24	0.7–0.9	< 0.1–0.1	< 0.1–0.1	0.4	0.5	< 0.1	0.2
25	0.7–0.9	< 0.1–0.1	< 0.1–0.1	0.4	0.5	< 0.1	0.2
26	0.7–1.0	< 0.1–0.1	< 0.1–0.1	0.4	0.5	< 0.1	0.2
27	0.8–1.0	0.1–0.2	< 0.1–0.1	0.4	0.5	< 0.1	0.2
28	0.8–1.1	0.1–0.2	< 0.1–0.2	0.4	0.5	< 0.1	0.2
29	0.8–1.2	0.1–0.2	< 0.1–0.2	0.5	0.5	< 0.1	0.2

30	0.9–1.2	0.1–0.2	<0.1–0.2	0.5	0.5	<0.1	0.2
31	0.9–1.3	0.1–0.2	<0.1–0.2	0.5	0.5	<0.1	0.2
32	1.1–1.5	0.1–0.2	0.1–0.2	0.6	0.5	<0.1	0.2
33	1.4–1.9	0.1–0.3	0.1–0.2	0.7	0.5	<0.1	0.2
34	1.9–2.4	0.2–0.4	0.1–0.3	0.7	0.5	<0.1	0.2
35	2.5–3.9	0.3–0.5	0.2–0.3	0.9	0.5	<0.1	0.3
36	3.2–5.0	0.3–0.6	0.2–0.4	1.0	0.5	<0.1	0.3
37	4.1–6.4	0.4–0.7	0.2–0.5	1.1	0.5	<0.1	0.3
38	5.2–8.1	0.5–0.9	0.3–0.7	1.3	0.5	<0.1	0.3
39	6.6–10.5	0.7–1.2	0.4–0.8	1.5	0.5	<0.1	0.3
40	8.5–13.7	0.9–1.6	0.5–1.1	1.8	0.5	<0.1	0.3
41	10.8–17.9	1.1–2.1	0.6–1.4	2.2	0.5	<0.1	0.3
42	13.8–23.4	1.4–2.7	0.7–1.8	2.7	0.5	<0.1	0.3
43	17.6–30.6	1.8–3.5	0.9–2.4	3.3	0.5	<0.1	0.3
44	22.5–40.0	2.3–4.6	1.2–3.1	4.1	0.5	<0.1	0.3
45	28.7–52.3	2.9–6.0	1.5–4.1	5.1	0.5	<0.1	0.3
46	36.6–68.3	3.7–7.9	1.9–5.3	6.4	0.5	<0.1	0.3
47	46.6–89.3	4.7–10.3	2.4–6.9	8.2	0.5	<0.1	0.3
48	59.5–116.8	6.0–13.5	3.0–9.0	10.6	0.5	<0.1	0.3
49	75.8–152.7	7.6–17.6	3.8–11.8	13.8	0.5	<0.1	0.3

[a] XXX is excluded.
[b] Calculation of the total at each age assumes rate for autosomal aneuploidies is at the midpoints of the ranges given.
[c] No range may be constructed for those under 20 years by the same methods as for those 20 and over.

Reprinted from ref. 13 with permission from The American College of Obstetricians and Gynecologists (*Obstetrics and Gynecology* (1981), **58**, p. 284).

94 *Human Cytogenetics*

Table 2. Maternal-age-specific risks for trisomy 21

MA[a]	Incidence at CVS		Incidence at AC		Incidence at LB	
	%	1 in	%	1 in	%	1 in
35	0.4	240	0.4	260	0.25	380
36	0.6	175	0.5	200	0.35	290
37	0.75	130	0.6	160	0.4	230
38	1.0	100	0.8	120	0.6	180
39	1.3	75	1.0	100	0.7	140
40	1.8	55	1.3	75	0.9	110
41	2.4	40	1.7	60	1.2	80
42	3.2	30	2.2	45	1.5	65
43	4.2	25	2.8	35	2.0	50
44	5.6	18	3.6	30	2.5	40
45	7.5	13	4.5	20	3.2	30
46	10	10	5.8	17	4.0	25
47	13	7				
48	18	6	2.3	45	1.6	60
49						

AC, amniocentesis; CVS, chorionic villus sampling; LB, live birth; MA, maternal age.

[a]Figures apply to midpoint of a maternal age interval.

Data derived from ref. 14, © John Wiley & Sons Ltd. Reprinted by permission of John Wiley & Sons Ltd.

Table 4. Maternal-age-specific risks for chromosomal abnormalities at chorionic villus sampling (excluding trisomy 21)

MA	Nonlethal abnormalities[a]		Lethal abnormalities[b]		Total	
	%	1 in	%	1 in	%	1 in
35	0.4	290	0.15	670	0.9	110
36	0.5	220	0.2	550	1.2	80
37	0.6	160	0.2	450	1.6	65
38	0.8	125	0.25	380	2.1	50
39	1.0	95	0.3	300	2.7	35
40	1.4	75	0.4	260	3.6	30
41	1.8	55	0.5	210	4.7	20
42	2.4	40	0.55	180	6.1	16
43	3.1	35	0.65	150	8.0	12
44	4.1	25	0.75	130	11	10
45	5.3	19	0.9	110	14	7
46	7.0	14	1.0	100	18	6
47	9.2	11	1.1	90	24	4
48	12.0	8	1.2	80	30	3

[a]Mostly +13, +18, XXX, XXY, 45X.

[b]Mostly nonviable trisomies and triploidy.

Data derived from ref. 14 with permission from Oxford University Press.

Table 3. Maternal-age-specific risks for trisomies 18 and 13

	Incidence at amniocentesis				Incidence at live birth			
	Trisomy 18		Trisomy 13		Trisomy 18		Trisomy 13	
MA	%	1 in	%	1 in	%	1 in	%	1 in
35	0.05	2000	0.02	5000	0.02	6000	0.01	9000
36	0.07	1400	0.03	3300	0.02	4500	0.02	6000
37	0.10	1000	0.04	2500	0.03	3000	0.02	4500
38	0.14	700	0.05	2000	0.05	2000	0.03	3500
39	0.2	500	0.08	1200	0.06	1600	0.05	2000
40	0.3	350	0.1	900	0.09	1100	0.06	1600
41	0.4	250	0.15	700	0.13	800	0.09	1200
42	0.55	180	0.2	500	0.2	600	0.01	800
43	0.75	130	0.05	2000	0.25	400	0.03	3500
44–49	0.5	200	0.05	2000	0.15	650	0.03	3500

Data reproduced from ref. 14 with permission from Oxford University Press.

Table 5. Maternal serum screening for Down syndrome using age, AFP, uE_3 and hCG in combination: detection rate (DR) and false-positive rate (FPR) in different maternal age groups, according to cutoff[a]

	Risk cutoff					
	1:200		1:250		1:300	
MA at EDD	DR (%)	FPR (%)	DR (%)	FPR (%)	DR (%)	FPR (%)
>35	83	19	86	22	88	26
<35	43	2.7	48	3.6	52	4.6
>38	90	31	92	36	94	40
<38	47	3.2	52	4.2	56	5.2
>40	94	42	96	48	96	53
<40	50	3.5	54	4.5	58	5.6
Any	57	3.9	61	5.0	64	6.1

[a] A result is positive if the risk of a Down syndrome pregnancy based on age, AFP, uE_3 and hCG is greater than, or equal to, the specified risk cutoff.
MA at EDD, maternal age at estimated date of delivery.
Data reproduced from ref. 15 with permission from Butterworth-Heinemann Ltd.

Table 6. Serum screening for Down syndrome: detection rate (DR), false positive rate (FPR) and odds of being affected given a positive screening result (OAPR) using maternal age, AFP, uE$_3$ and hCG according to risk cutoff level, method of estimating gestational age (dates or scan) and maternal weight (unadjusted or adjusted)[a]

Risk cutoff	Dates and unadjusted for weight			Dates and adjusted for weight			Scan and unadjusted for weight			Scan and adjusted for weight		
	DR (%)	FPR (%)	OAPR	DR (%)	FPR (%)	OAPR	DR (%)	FPR (%)	OAPR	DR (%)	FPR (%)	OAPR
1:150	48	2.7	1:44	49	2.8	1:43	57	2.6	1:35	58	2.6	1:34
1:200	54	3.9	1:56	54	4.0	1:56	62	3.6	1:44	62	3.6	1:44
1:250	58	5.2	1:68	59	5.2	1:68	65	4.6	1:54	66	4.6	1:53
1:300	62	6.4	1:80	62	6.4	1:79	68	5.6	1:63	69	5.5	1:62
1:350	65	7.6	1:91	65	7.6	1:90	71	6.6	1:71	71	6.5	1:71

[a]A result is positive if the risk of a Down syndrome pregnancy, given the maternal age and the three marker levels, is greater than or equal to the specified cutoff risk.

Data reproduced from ref. 16 with permission from Churchill Livingstone.

Table 7. Chromosome abnormalities associated with structural abnormalities in the fetus

Fetal abnormality on scan	Chromosome abnormalities
Nonimmune hydrops, edema, ascites	45, X
Cystic hygromata	45, X, Trisomy 21
Increased nuchal skin thickness	Trisomy 21, 45, X
Abnormal head shape, doliocephaly	Trisomy 18; 21
Choroid plexus cysts	Trisomy 18
Agenesis of corpus callosum	Trisomy 18
Facial clefting	Trisomy 13, 4p-
Duodenal atresia (double bubble)	Trisomy 21
Omphalocele	Trisomy 13; 18
Cardiovascular defects, e.g. ASD/VSD	Trisomy 13; 18; 21 Triploidy
Pleural effusion	Trisomy 21
Renal anomalies, particularly obstructive uropathy	Trisomy 18; 13
Syndactyly/polydactyly	Triploidy; Trisomy 13
Aplasia thumb/radius	Fanconi's syndrome, 13q
Clasped, overlapping fingers	Trisomy 18

ASD, atrial septal defect; VSD, ventricular septal defect.
Data derived from ref. 17 with permission from Butterworth-Heinemann Ltd.

Table 8a. Definitions of mosaicism in CVS and amniotic fluid

Level	Cell number	Distribution
I	Single	Any culture or CVS direct preparation
II	Multiple, same abnormality	Single colony or flask of amniotic fluid cell, or long-term CVS cultures
III	Multiple, same abnormality	Multiple colonies or flasks of amniotic fluid cell or long-term CVS cultures. Also CVS direct preparation, discordant nonmosaic abnormality in cells from either CVS direct preparations or long-term cultures

Table 8b. CVS level III mosaicism

	Type A	Type B	Type C
CVS direct preparation	Abnormal	Normal	Abnormal
CVS long-term culture	Normal	Abnormal	Abnormal

In type C, at least one of the results must be mosaic.
Abnormal refers to mosaic or nonmosaic abnormal karyotype.

Table 9. Potential significance and value of chromosome results from chorionic villi

Reliable indicators of the fetal karyotype:
  nonmosaic trisomy 13, 18 or 21
  nonmosaic numerical sex chromosome abnormality if present in
    both direct preparations and cultured cells
  triploidy
  the outcome of known familial rearrangements
  the sex as determined from direct preparations

Unreliable indicators of the fetal karyotype:
  mosaic trisomy 13, 18 or 21
  de novo supernumerary markers
  de novo nonmosaic structural abnormalities if only a result from
    direct preparations is available
  mosaic sex chromosome abnormalities
  45, X in direct preparations

Poor indicators of the fetal karyotype:
  mosaic or nonmosaic autosomal trisomies other than
    chromosomes 13, 18 or 21
  mosaic structural rearrangements
  tetraploidy

Data reproduced from ref. 20, © 1994 John Wiley & Sons Ltd. Reprinted by permission of John Wiley & Sons Ltd.

Table 10. Percentage mosaicism excluded with 0.90, 0.95 and 0.99 confidence if specified number of cells are evaluated and found to have identical karyotypes

No. of cells (*n*)	Confidence levels (%)		
	0.90	0.95	0.99
6	32	40	54
10	21	26	37
15	15	19	27
20	11	14	21
25	9	12	17
30	8	10	15
50–55	5	6	9
59–63	4	5	8
99–112	3	3	5
152–227	2	2	3
299–458	1	1	2

If *n* = the number of cells counted, then the degree of some variant line not observed excluded with 90%, 95% or 99% confidence is given in the appropriate column. (If cells are grown in long-term culture, then the number of independent colonies should be used.) For example, if one wants to exclude a level of 21% or greater of some abnormal chromosome line in a tissue sampled with 99% confidence, then one must evaluate 20 independent cells (and find the same normal pattern in each).
Data derived from ref. 22 with permission from Churchill Livingstone.

Table 11. Assessment of gestational age in weeks plus days from the crown–rump length (CRL)

CRL (mm)	Percentile*			CRL (mm)	Percentile		
	5th	50th	95th		5th	50th	95th
10	6 + 5	7 + 3	8	30	9 + 5	10 + 2	11
11	6 + 6	7 + 4	8 + 2	31	9 + 5	10 + 3	11 + 1
12	7 + 1	7 + 5	8 + 3	32	9 + 6	10 + 4	11 + 2
13	7 + 2	8	8 + 4	33	10	10 + 5	11 + 2
14	7 + 3	8 + 4	8 + 6	34	10 + 1	10 + 6	11 + 3
15	7 + 4	8 + 2	9	35	10 + 2	10 + 6	11 + 4
16	7 + 5	8 + 3	9 + 1	36	10 + 2	11	11 + 5
17	8	8 + 4	9 + 2	37	10 + 3	11 + 1	11 + 6
18	8 + 1	8 + 5	9 + 3	38	10 + 4	11 + 2	11 + 6
19	8 + 2	8 + 6	9 + 4	39	10 + 5	11 + 2	12
20	8 + 3	9	9 + 5	40	10 + 5	11 + 3	12 + 1
21	8 + 4	9 + 1	9 + 6	41	10 + 6	11 + 4	12 + 1
22	8 + 5	9 + 2	10	42	11	11 + 4	12 + 2
23	8 + 6	9 + 3	10 + 1	43	11	11 + 5	12 + 3
24	8 + 6	9 + 4	10 + 2	44	11 + 1	11 + 6	12 + 3
25	9	9 + 5	10 + 3	45	11 + 2	11 + 6	12 + 4
26	9 + 1	9 + 6	10 + 4	46	11 + 2	12	12 + 5
27	9 + 2	10	10 + 5	47	11 + 3	12 + 1	12 + 5
28	9 + 3	10 + 1	10 + 5	48	11 + 4	12 + 1	12 + 6
29	9 + 4	10 + 2	10 + 6	49	11 + 4	12 + 2	13

*Week of gestational age (weeks plus days) from the crown–rump length.
Reproduced from ref. 23 with permission from Blackwell Scientific Publications.

Table 12. Average predicted fetal measurements at specific menstrual age

Menstrual age (weeks)	BPD (cm)	Head circum. (cm)	Femur length (cm)
12.0	1.7	6.8	0.7
13.0	2.1	8.2	1.1
14.0	2.5	9.7	1.4
15.0	2.9	11.0	1.7
16.0	3.2	12.4	2.0
17.0	3.5	13.8	2.4
18.0	3.9	15.1	2.7
19.0	4.3	16.4	3.0
20.0	4.6	17.7	3.3
21.0	5.0	18.9	3.5
22.0	5.3	20.1	3.8
23.0	5.6	21.3	4.1
24.0	5.9	22.4	4.4
25.0	6.2	23.5	4.6
26.0	6.5	24.6	4.9
27.0	6.8	25.6	5.1
28.0	7.1	26.6	5.4
29.0	7.3	27.5	5.6
30.0	7.6	28.4	5.8
31.0	7.8	29.3	6.0
32.0	8.1	30.1	6.2

33.0	8.3	30.8	6.4
34.0	8.5	31.5	6.6
35.0	8.7	32.2	6.8
36.0	8.9	32.8	7.0
37.0	9.0	33.3	7.2
38.0	9.2	33.8	7.4
39.0	9.3	34.2	7.5
40.0	9.4	34.6	7.7

Derived from ref. 23.

Chapter 7 THE ROLE OF CYTOGENETICS IN THE INVESTIGATION OF MUTAGEN EXPOSURE AND CHROMOSOME INSTABILITY – E.J. Tawn and D. Holdsworth

The study of chromosome aberrations has wide applications. A number of endpoints can be examined depending on the test situation. In order to observe the following endpoints it is necessary to induce one or more cycles of cell division. Chromosome aberrations and sister chromatid exchanges manifest themselves in cells at metaphase, whereas micronuclei are detectable in interphase cells. Full culturing procedures and scoring criteria have been described previously [1]. It is only possible here to present a brief outline and we recommend referral to more detailed texts before embarking on this type of work.

1.3 Sister chromatid exchange (SCE)

1. Reciprocal exchanges between sister chromatids.
2. Detection requires two division cycles in the presence of bromodeoxyuridine, which allows differential staining of chromatids by fluorescence plus Giemsa.

1.4 Micronuclei

1. Discrete round bodies of nuclear origin observed in the interphase cytoplasm.
2. Derived from acentric fragments or chromosomes which have lagged at mitosis.

1 Description of cytogenetic endpoints

1.1 Chromosome type aberrations

1. Breaks and exchanges involve both chromatids at identical loci.
2. Cells must undergo a cycle of DNA replication to allow expression.
3. Classified as symmetrical (e.g. translocations, inversions) or asymmetrical (e.g. dicentrics, rings).
4. Asymmetrical aberrations easily identified with conventional block staining (*Figure 1*).
5. Identification of symmetrical aberrations usually requires banding or fluorescence *in situ* hybridization (FISH).

1.2 Chromatid type aberrations

1. Breaks and exchanges involve one chromatid only.
2. Configurations produced allow most types of both asymmetrical and symmetrical aberrations to be detected using block staining (*Figure 1*).
3. Some will give rise to derived chromosome type aberrations following further cell division.
4. Cells which have undergone division can be identified by their binucleate appearance following addition of cytochalasin-B.

2 Applications

Chromosome aberration analysis can be employed for radiation dose assessment [3, 4], population monitoring of genotoxic exposure [5, 6], *in vitro* chemical testing [7, 8, 9] and in the investigation of chromosome instability syndromes [10].

2.1 Radiation dose assessment

Chromosome analysis of peripheral blood lymphocytes for the assessment of radiation exposure provides a means of validating physical dosimetry and the opportunity to estimate dose when no other measurements are available. Historically this has only been applicable to recent acute doses, but advances in methodology now offer the potential to assess the extent of past and chronic exposures. Essential features of the technique are outlined overleaf.

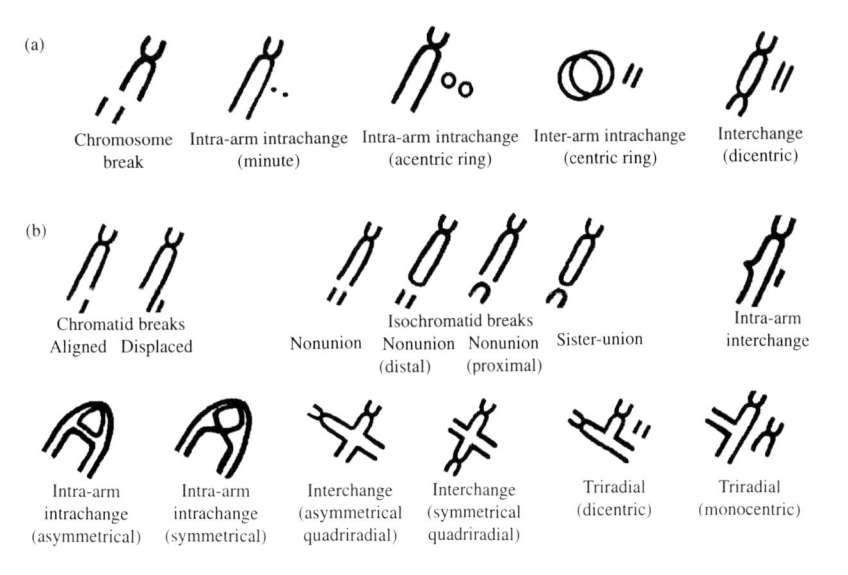

Figure 1. The principal forms of chromosome damage easily identified by block staining. (a) Chromosome type; (b) chromatid type. Reproduced from ref. 2 with permission from Oxford University Press.

1. Chromosome type aberrations in first division lymphocyte cultures.
2. Dicentric frequency employed for estimation of recent acute accidental whole-body dose > 50 mSv. Sampling undertaken between 24 h and 1 month post-exposure.
3. Translocation frequencies determined by FISH used to estimate accumulated chronic exposures [4]. This technique is still under evaluation.
4. Identification of potential confounders, for example smoking.
5. Refer to *in vitro* dose–response curves of appropriate radiation quality.

2.2 Population monitoring

Increasing concern about the genotoxic effects of a wide range of environmental agents has resulted in a need to develop and apply methods of studying somatic and germinal genetic changes on a population basis. The type and degree of cytogenetic damage observed will be determined by the nature of the agent. Population monitoring is both time consuming and expensive and therefore needs careful planning. Extensive guidelines for cytogenetic studies are available [5, 6] and only the key points that need to be considered are described here.

1. Knowledge of nature and/or source of clastogen.
2. Prior demonstration of effect experimentally.
3. Identification of exposed population.
4. Identification of suitable unexposed control group.
5. Measure of exposure, for example duration, physical, biological.
6. Separation into several exposure levels.
7. Knowledge of potential confounders.

2.3 *In vitro* chemical testing

Cytogenetic investigations form part of the wide range of tests which are routinely employed in the investigation of the potentially mutagenic effects of chemicals. The following highlights some of the key points to consider if these techniques are contemplated, but we strongly recommend a more detailed review of the concepts and practicalities of genetic toxicology testing [7, 8, 9].

1. Knowledge of physical and chemical properties of test substance.

2. Preliminary cytotoxicity testing to derive appropriate concentration range for subsequent testing.
3. Some chemicals may require metabolic activation to allow expression of mutagenic effect.
4. Use blood from two or more donors with no recent history of smoking or medication.
5. Time and duration of treatment dependent upon nature of substance and its interaction with other components of culture system.
6. Cultures to control all aspects of the test protocol should be employed, for example activation solvent.

3 Chromosome instability syndromes

The chromosome instability syndromes comprise a number of rare but well-defined recessively inherited clinical entities. In the main they are disorders characterized by defective DNA repair or replication. Although the majority are distinct, some disorders overlap which may reflect a similar

3.1 Observed cytogenetic changes
Ataxia telangiectasia

1. Spontaneous translocations and inversions in T lymphocytes involving immunoglobulin and T-cell receptor genes particularly at 7p13, 7q35, 14q11, 14q32 and, less frequently, 2 and 22.
2. Increased frequencies of chromatid type aberrations induced in G_0/G_1 and notably G_2 lymphocytes following treatment with X-irradiation or bleomycin.

Bloom's syndrome

1. 10–15-fold increase in spontaneous SCE frequencies.
2. Elevated levels of spontaneous chromatid aberrations, particularly symmetrical quadriradials.
3. Some evidence for increased sensitivity to alkylating agents.

Fanconi's anemia

1. Elevated spontaneous levels of chromatid and derived

underlying mechanism, an example being Nijmegen breakage syndrome which shows a number of features in common with ataxia telangiectasia.

Xeroderma pigmentosum (XP) exhibits extensive heterogeneity and overlaps with Cockayne's syndrome (CS) and trichothiodystrophy (TTD). Complementation studies have identified seven different genes involved in the primary defect in XP. In the classical form of CS two complementation groups have been identified but some patients have been assigned to XP group B and some to XP group D. In addition, a number of cases of TTD have been assigned to XP group D. These overlaps may be explained by the multiple functions of repair proteins. Different viable mutations may have a range of different systemic effects, and each disease may represent a different protein–protein interaction.

Table 1 summarizes the major clinical features of these syndromes.

chromosome aberrations.
2. Marked increase in induced levels of chromatid aberrations following treatment with cross-linking agents, for example mitomycin C or diepoxybutane.

ICF syndrome (immunodeficiency, centromeric heterochromatin instability and facial anomalies)

1. Morphological abnormalities of heterochromatic centomeric regions of chromosomes 1, 16 and, to a lesser extent, 9 leading to associations, whole-arm deletions and multibranched configurations in lymphocytes.
2. Elevated response with respect to all the above, plus other chromosome abnormalities following treatment with mitomycin C.

Robert's syndrome

1. Distinctive premature separation of centromeric heterochromatic regions in metaphase chromosomes, particularly 1, 9, 16 and also Yq.

2. Chromosome lagging at anaphase.

Werner's syndrome

1. Variegated translocation mosaicism (stable clonal rearrangements) in fibroblast cultures.
2. Slightly increased levels of chromatid aberrations may or may not be observed in lymphocytes.

Xeroderma pigmentosum

1. Increased induced levels of SCE and aberrations (mostly chromatid) following exposure to ultraviolet (UV) light.

2. Low level of UV-induced unscheduled DNA synthesis.

Cockayne's syndrome
Similar cytogenetic sensitivity to XP.

Trichothiodystrophy
A subset of this disorder exhibits abnormal photosensitivity similar to XP.

Table 2 summarizes the spontaneous and induced cytogenetic changes that occur in these syndromes.

Table 1. Major clinical features and patterns of inheritance

Disease	Major clinical feature	Frequency	Ref.
Ataxia telangiectasia	Progressive cerebellar ataxia, telangiectases of conjunctiva and skin, immunodeficiency, predisposition to lymphoid malignancies	1:100 000–1:300 000[a]	11,12
Bloom's syndrome	Telangiectatic erythema, short stature, mental retardation in some cases, immunodeficiency, predisposition to malignancies especially acute leukemia	Rare[b]	13,14
Fanconi's anemia	Progressive pancytopenia, *café-au-lait* spots, skeletal abnormalities, short stature, predisposition to malignancy especially AML	1:300 000	15
Nijmegen breakage syndrome	Short stature, microcephaly, immunodeficiency	Rare	16
ICF syndrome	Immunodeficiency, mild developmental delay, facial abnormalities	Rare	17
Robert's syndrome	Hypomelia, mid facial defect, severe growth deficiency	Rare	18
Werner's syndrome	Premature ageing, short stature, predisposition to malignancy particularly sarcomas	Rare	19
Xeroderma pigmentosum	Photosensitivity, high incidence of (predominantly skin) neoplasia, progressive neurological abnormalities	1:250 000[c] 1:40 000[d]	20
Cockayne's syndrome	Dwarfism, mental retardation, progressive neurological degeneration, photosensitivity	Rare	21,22
Trichothiodystrophy	Brittle hair (defective sulfur metabolism), ichthyosis, mental and physical retardation, photosensitivity in some cases	Rare	22

[a] Considerable ethnic variation.
[b] Approx. 130 cases identified world-wide – half in Ashkenazi Jews.
[c] USA and Europe.
[d] Japan.

Table 2. Summary of spontaneous and induced cytogenetic changes in chromosome instability syndromes

Syndrome	Spontaneous		Induced		
	Chromosome abnormalities	SCE	Chromosome abnormalities	SCE	Inducing agent
Ataxia telangiectasia	+	N	+ +	N	X-rays, bleomycin
Bloom's syndrome	+ +	+ +	+	+	Alkylating agents ? radiation
Fanconi's anemia	+	N	+ +	N/+	DNA cross-linking agents
Nijmegen breakage syndrome	+	N	+ +	N	X-rays, bleomycin
ICF syndrome	+	N	+ +	ns	Mitomycin C
Robert's syndrome	+	ns	ns	ns	
Werner's syndrome	+	N	+	N	Diepoxybutane 4-nitroquinoline oxide
Xeroderma pigmentosum	N	N	+ +	+	UV
Cockayne's syndrome	N	N	+ +	+	UV
Trichothiodystrophy	N	N	+ +	+	UV

N, normal response; +, elevated response; + +, markedly elevated response; ns, not studied.

Chapter 8 **HEALTH AND SAFETY DATA** – R.T. Howell and S.H. Roberts

1 General data

1.1 Management responsibility to staff

1. Provide a safe working environment.
2. Make a suitable and sufficient assessment of risk to health arising out of all work activities.
3. Take effective control measures to reduce risks.
4. Provide training.
5. Provide written systems of work/codes of practice.
6. Provide protective clothing and safety equipment.
7. Maintain equipment and workplace on a regular basis.
8. Provide first aid facilities.
9. Offer inoculation against hepatitis B, tuberculosis, and tetanus for all laboratory staff.

1.2 Staff responsibility

1. Use safety equipment and protective clothing.
2. Follow protocols and local rules for safe use of equipment.
3. Inform Safety Advisor of change in system of work.
4. Not to put themselves or others at risk.

1.3 Sources of expert advice

1. Safety Officer of the Institution.
2. Hospital Occupational Health Department.
3. Safety Advisor of the Association of Clinical Cytogeneticists.
4. Health and Safety Executive (HSE).

112 *Human Cytogenetics*

1.4 Health and safety organization within the laboratory

1. Safety Advisor.
2. Radiological Protection Supervisor (if handling radio-active substances).
3. Safety Committee meeting regularly.
4. Accident book to record all accidents and potentially hazardous occurrences.
5. Regular safety checks undertaken by Safety Advisor or deputy.
6. Maintenance of records to comply with statutory regulations.
7. Staff health records to identify any work-related patterns of sickness.

1.5 Protective clothing to be made available to staff handling cytotoxic and pathological materials

1. Laboratory overall (to be worn properly fastened);
2. Disposable plastic/latex gloves;
3. Face mask;
4. Avoid the use of sharps;
5. Protective clothing to include gloves and plastic apron;
6. Remove unnecessary equipment from work station;
7. Consult with other staff and deny access to unauthorized persons;
8. Batch-process high-risk samples;
9. Permissive cells (i.e. with CD4 receptor) such as lymphocytes, culture duration limited to 100 h.

2.2 Categories of high-risk sample
Hepatitis B

1. Patients in renal units;
2. Patients suffering from infective or suspected infective diseases of the liver;
3. Patients receiving immunosuppressive or cytotoxic drugs;
4. Patients with mental handicap who are, or have been, resident in an institution;
5. Drug abusers;
6. Homosexuals;

4. Disposable plastic apron;
5. Disposable plastic oversleeves;
6. Goggles/vizor.

2 Pathological hazards

2.1 Containment of pathological hazards

This is described in *Table 1*.

Basic requirements; containment levels 2 and 3
See ref. 2.

1. Suitable microbiological safety cabinet;
2. Autoclave available;
3. Wash-hand basin;
4. Partially isolated room (3 only);
5. Inward airflow (3 only);
6. Room sealable for fumigation (3 only).

Containment ('2 + ') for culturing the majority of samples in the cytogenetics laboratory

1. Delineated/secluded class 1 (preferably) or 2 safety cabinet;

7. Sexual partners of the above;
8. Children of seropositive mothers.

HIV high risk (1–3) and groups with increased prevalence of infection (4–7)

1. Confirmed or suspected cases of AIDS;
2. Patients with symptoms indicative of HIV infection, e.g. persistent generalized lymphadenopathy (PGL);
3. Persons known to have HIV antibody or to be positive for other laboratory markers for this infection;
4. Homosexual and bisexual males;
5. Intravenous drug users;
6. Hemophiliacs who have received untreated blood;
7. Persons having had sexual contact or transfusions in high-prevalence areas (e.g. sub-Saharan Africa).

3 Decontamination and waste disposal

3.1 Disinfectants and decontaminants in common use
(Important: refer to local policies and HSE guidelines.)

1. 70% Ethanol (not methanol), highly flammable;
2. 1% sodium hypochlorite freshly prepared, corrosive, do not mix with acid;
3. 'Virkon', 1%, or as per manufacturer's instructions, avoid prolonged contact with metals;
4. 2% glutaraldehyde, active 14 days, highly toxic;
5. 4% formaldehyde, formalin, fumigation of microbiological safety cabinets only.

Table 2 gives the activities of some common disinfectants.

3.2 Waste disposal

1. Adhere strictly to regulations and local arrangements.
2. Segregate clinical waste into approved color-coded bags.
3. Autoclave high-risk materials before disposal.
4. All sharps to go in rigid container, interior may be disinfected with 70% ethanol.
5. Decontaminated biological fluids (e.g. tissue culture products) via designated sink, flush well with water.
6. Stains; small volumes via sink, flush well with water.

2. Beware particularly of substances labeled R39, R40, R45 and R49 which are possible mutagens and carcinogens, and R46 and R47 which include mutagens and teratogens.
3. NB: the absence of R numbers does not necessarily imply that a substance is safe.

4.4 Classes of toxic substance commonly used in cytogenetics

1. *Alkylating agents* react with bases in the DNA, inducing gene mutations, chromosomal aberrations and sister chromatid exchange.
2. *Intercalators and DNA ligands* bind to the DNA, causing distortions and disrupting synthesis and condensation. Some are known to potentiate the effects of mutagenic agents.
3. *Nucleoside and base analogs* are incorporated into the DNA at synthesis in place of normal bases/nucleosides and can compromise the stability of DNA. They are prone to mispairing at S phase of the cell cycle thereby inducing base substitution mutations.

4 Chemical hazards

4.1 Chemicals assigned inhalation exposure limits

These chemicals are listed in *Table 3*. See also ref. 3.

4.2 Cytotoxic chemicals

1. Mutagens induce mutation.
2. Clastogens induce chromosomal breakage and are therefore inevitably mutagenic.
3. Carcinogens induce tumors. Many mutagens are proven carcinogens.
4. Teratogens affect fetal development and are thus responsible for birth defects.

4.3 Relevant risk phrases for hazardous chemicals used in cytogenetics

These are listed in *Table 4*. See also ref. 4.

1. R numbers separated by a '/' indicate a combination of particular risks.

4. *Antimetabolites* block DNA replication by inhibition of one or more steps in nucleotide synthesis. They enhance expression of fragile sites and are used as synchronizing agents to obtain elongated chromosomes.
5. *Transforming agents* stimulate the proliferation of normally quiescent cells. This in itself is not necessarily hazardous; however, tumorigenesis is associated with abnormal cell proliferation, and mutations are generated and fixed at cell division. The phorbol esters (e.g. phorbol 12-myristate 13-acetate (TPA)) are tumor promoters which enhance the effect of carcinogens.
6. *Mitotic spindle inhibitors* arrest cells at metaphase by binding to the protein tubulin of which the spindle fibers are composed. They can induce polyploidy and aneuploidy.

4.5 Uses of toxic substances in cytogenetics

1. *Stains and counterstains*. All chromosomal stains interact with DNA and/or associated proteins. Many are intercalators or base-pair-specific ligands.
2. *Condensation inhibition*. Intercalators and nucleoside

analogs are sometimes added to cultures to inhibit chromosome condensation (throughout or in specific regions).

3. *Replication studies*. The nucleoside analog 5′-bromodeoxyuridine added to cultures alters the staining properties of the fixed chromosomes.

4. *Synchronization*. Antimetabolites and the nucleoside analog fluorodeoxyuridine are used to synchronize cell cultures by blocking DNA synthesis.

5. *Fragile site induction*. The synchronizing agents which block DNA synthesis are also effective in inducing folate-sensitive fragile sites.

6. *Transformation*. Ingredients are added to stimulate mitotic activity of lymphocyte cultures.

7. *Mitotic block*. Addition of a blocking agent accumulates cells at metaphase.

8. *Aberration induction*. Chromosome damaging agents (clastogens) are used in the diagnosis of Fanconi's anemia and sometimes ataxia telangiectasia. These are generally bi- or polyfunctional alkylating agents.

9. In situ *hybridization*. This staining method employs various hazardous fluorochromes and formamide, a teratogen.

9. Keep solutions for freezing in strong bottles.

10. Clearly label all containers, including Coplin jars containing fluorochromes.

11. Dispense solutions using disposable plastic tips, pipettes and bottles as far as possible.

12. Discard used solutions as soon as possible.

13. Flush solutions other than organic solvents down the drain with water. Leave the tap running for some minutes.

14. Small volumes of toxic solutions may be absorbed on vermiculite/Fuller's earth in a plastic container such as a urine pot. Tape the lid securely and send for incineration.

15. If in doubt, consult the hospital pharmacy, occupational health or other appropriate expert.

16. When adding liquid to single-use vials designed for dilution by syringe and needle, always use a second 'vent needle' to avoid build up of pressure in the vial.

17. Cultures should be capped and centrifuged in sealed buckets, opening the buckets only under exhaust protection.

Table 5 lists cytotoxic chemicals used in cytogenetics.

4.6 Safe use of toxic substances

1. Wear suitable protective clothing.
2. Discard protective clothing in a logical order (e.g. remove apron before gloves, gloves before face mask).
3. As far as possible, carry out procedures in exhaust protective cabinet (fume cupboard or appropriate tissue culture cabinet).
4. Avoid weighing if possible. Purchase powders in small pre-weighed amounts.
5. Weigh in stoppered glass bottles, not plastic weighing boats because of static.
6. Make up stock solutions and dilution series in small volumes. This way spillages are more easily contained.
7. Minimize risk of spillage by keeping bottles in racks, and handle in a tray and/or on 'Benchkote'.
8. Keep a spillage absorption material available, for example Fuller's earth or vermiculite.

5 Publications concerning statutory instruments, Health and Safety Executive guidelines and EC directives

These publications are applicable to work carried out in the United Kingdom. For work in other countries, consult your local Health and Safety Office.

5.1 General
Health and Safety at Work etc. Act (1974) HMSO, London.
A Guide to the Reporting of Injuries, Diseases, and Dangerous Occurrences Regulations (1985). HMSO, London.
HSAC, Guidance on the Recording of Accidents and Incidents in the Health Services (1986). HMSO, London.
Pregnant and Breast Feeding Workers Directive 92/85/EEC (1992). HMSO, London.
Personal Protective Equipment at Work; Guidance on Regulations (1992) HMSO, London.
HSE (1993). *Management of Health and Safety at Work, Approved Code of Practice*. HMSO, London.

SI (1985) *The Ionising Radiations Regulations*, SI 1985 no. 1333. HMSO, London.

HSE (1993) *Work Equipment, Guidance on Regulations*. HMSO, London.

HSE (1993) *Manual Handling, Guidance on Regulations*. HMSO, London.

HSE (1993) *Workplace Health, Safety and Welfare*, Approved Code of Practice. HMSO, London.

HSE (1993) *Display Screen Equipment Work, Guidance on Regulations*. HMSO, London.

5.2 Pathogens

ACDP (1990) *Categorisation of Pathogens According to* Hazard and Categories of Containment. HMSO, London.

ACDP (1990) *HIV – the Causative Agent of AIDS and Related Conditions. Second Revision of Guidelines*. HMSO, London.

Safe Working and the Prevention of Infection in Clinical Laboratories (1991). HMSO, London.

Safe Working and the Prevention of Infection in Clinical Laboratories – Model Rules for Staff and Visitors (1991). HMSO, London.

5.3 Chemicals

HSE, Guidance Note EH40, *Occupational Exposure Limits*, revised annually. HMSO, London.

Control of Substances Hazardous to Health (general ACoP) and *Control of Carcinogenic Substances* (carcinogens ACoP) (1988). HMSO, London.

SI (1993) *The Chemicals (Hazard Information and Packaging)* Regulations, no. 1746. HMSO, London.

HSE (1993) *CHIP for Everyone (HSG 108)*. HMSO, London.

HSE (1993) *The Approved Guide to the Classification and Labelling of Substances and Preparations Dangerous for Supply (L38)*. HMSO, London.

HSE (1993) *Information Approved for the Classification and Labelling of Substances and Preparations Dangerous for Supply*. HMSO, London.

Table 1. Containment of pathological hazards

Pathogen	Containment level[a]
Hepatitis B + delta	2+
HIV	3
Non-A, non-B hepatitis, including C	2
Epstein–Barr virus	2
EB virus LCL continuous culture	3
Cytomegalovirus	2
CD4+ permissive cells (<100 h culture)	2+
CD4+ permissive cells (>100 h culture)	3
Herpes	2
Syphilis	2
Tuberculosis	3
Malaria	4
Rubella	2
Rubeola	2
Toxoplasmosis	3
Listeria monocytogenes	2

[a]For definitions of containment categories see ref. 1.
LBL, lymphoblastoid cell line.

Table 2. Activities of some common classes of disinfectant

Disinfectant	Active against				
	Vegetative bacteria	Bacterial spores	Fungi	Lipid viruses	Nonlipid viruses
Hypochlorite	+	+	1	+	+
Phenolic	+	−	+	+	2
Alcohols	+	−	−	+	2
Aldehydes	+	+	+	+	+
Surfactants	+	−	1	2	2

1, Limited fungal activity; 2, dependent on the virus; +, generally effective; −, not reliably effective.

Table 3. Chemicals assigned inhalation exposure limits

Chemical	8 h TWA	10 m MEL
Acetic acid	10 p.p.m.	15 p.p.m.
Ethanediol (in Giemsa, etc.)	10 mg m^{-3}	
Ethanol	1000 p.p.m.	
Formamide	20 p.p.m.	30 p.p.m.
Glutaraldehyde	0.2 p.p.m.	0.2 p.p.m.
Methanol	200 p.p.m.	250 p.p.m.
Xylene (80% in DPX)	100 p.p.m.	150 p.p.m.

MEL, maximum exposure limit; TWA, time-weighted average.

Table 4. Risk phrases for hazardous chemicals used in cytogenetics

R10	Flammable
R11	Highly flammable
R20	Harmful by inhalation
R21	Harmful in contact with skin
R22	Harmful if swallowed
R23	Toxic by inhalation
R24	Toxic in contact with skin
R25	Toxic if swallowed
R26	Very toxic by inhalation
R27	Very toxic in contact with skin
R28	Very toxic if swallowed
R36	Irritating to eyes
R37	Irritating to respiratory system
R38	Irritating to skin
R39	Danger of very serious irreversible effects
R40	Possible risk of irreversible effects
R41	Risk of serious damage to eyes
R42	May cause sensitization by inhalation
R43	May cause sensitization by skin contact
R45	May cause cancer
R46	May cause heritable genetic damage
R47	May cause birth defects
R49	May cause cancer by inhalation

Table 5. Cytotoxic chemicals used in cytogenetics

Substance	Use	R number
Acetic acid	Fixative	10,35
Acridine orange	Stain	20/21/22,40
Actinomycin D	Counterstain anti-condensing agent	26/27/28,36,37,38,47
Adriamycin	Clastogen	23/24/25,36/37/38,45,46
Amethopterin	See methotrexate	
5'-Azacytidine	Anti-condensing agent	20/21/22,45,46
Bleomycin	Clastogen	26/27/28,45,46
5'-Bromodeoxyuridine	Replication studies	20/21/22,46,47
Chromomycin A3	Stain	
Colcemid	Mitotic block	23/24/25,40
Colchicine	Mitotic block	26/28
Cytochalasin B	Cytokinesis inhibitor	26/27/28,40
4,6-Diamidino-2-phenylindole (DAPI)	Stain	
Diepoxybutane	Clastogen	23/24/25,36/37/38,40,42/43
Distamycin A	Counterstain	
Doxorubicin	See adriamycin	
Ethidium bromide	Stain, anti-condensing agent	36/37/38,46
Fluorescein isothiocyanate (FITC)	Stain	20/21/22,36/37/38,42/43
5'-Fluorodeoxyuridine	Synchronization, fragile sites	23/24/25,40
Formamide	FISH	20/21/22,36/37/38,47
Giemsa	Stain	20/21/22,40,41
Hoechst 33258	Stain	40

Continued

Table 5. Cytotoxic chemicals used in cytogenetics, *continued*

Substance	Use	R number
Leishman	Stain	See Giemsa
Methanol	Fixation, solvent for stains	11,23/25
Methotrexate	Synchronization, fragile sites	23/24/25,36/37/38,46,47
Methyl Green	Counterstain	36/37/38
Mitomycin C	Clastogen	26/27/28,45,46
Mustine hydrochloride	Clastogen	26/27/28,45,46,47
Nitrogen mustard	See mustine	
Olivomycin	Stain	
Phorbol 12-myristate 13-acetate (TPA)	Mitogen	45
Phytohemagglutinin (PHA)	Mitogen	
Podophyllotoxin	Mitotic block	
Pokeweed mitogen	Mitogen	
Propidium iodide	Stain	36/37/38,46
Quinacrine dihydrochloride	Stain	20/21/22
Quinacrine mustard	Stain	23/24/25,40,42/43
Thymidine	Synchronization, fragile sites	20/21/22,40
Vinblastine	Mitotic block	23/24/25,37/38,41,47
Vincristine	Mitotic block	23/24/25,36/37/38,47

Chapter 9 MANUFACTURERS AND SUPPLIERS

The following list of manufacturers and suppliers is, of neccessity, by no means exhaustive. However, the few companies included between them span all aspects of cytogenetic and associated molecular genetic technology. The prefixes refer to the principal types of product obtainable from the company (though prefixes do not cover the entire range of products that a company may offer).

The numbers bracketed are area, freephones or freefax numbers; international dialling codes have not been listed. UK (0800) and USA (800) freephone or freefax numbers can only be used in the corresponding countries.

Key to prefixes:
 C = chemicals
 CIS = computerized imaging systems
 E = laboratory equipment
 MIC = microscopes
 MG = products for molecular genetics
 P = photographic
 TCP = cell and tissue culture requisites, particularly
 TCR = cell and tissue culture media and reagents

(P) Agfa Gevaert Ltd,
27 Great West Road, Brentford TW8 9HN, UK

(C) Aldrich Chemical Co. Ltd,
The Old Brickyard, New Road, Gillingham, Dorset SP8 4JL, UK

US Address: Aldrich Chemical Co. Inc., PO Box 355, 1001 West St Paul Avenue, Milwaukee, WI 53233, USA. Tel. (414) 273 3850, (800) 558 9160; fax (800) 962 959

(MG, CIS) Alpha Laboratories Ltd,
40 Parham Drive, Eastleigh, Hants SO5 5ZU, UK
Tel. (0703) 610911
Fax (0703) 643701

124 *Human Cytogenetics*

(MG) **Amersham International plc.**
Amersham Place, Little Chalfont, Bucks HP7 9NA, UK
Tel. (0296) 395222, (0800) 515313
Fax (0296) 85910

US Address: Amersham Corp., 2636 South Clearbrook Drive, Arlington Heights, IL 60005, USA. Tel. (800) 323 9750; fax (708) 437 1640

(MG) **Applied Biosystems Ltd,** Kelvin Close, Birchwood Science Park North, Warrington, Cheshire WA3 7PB, UK
Tel. (0925) 825650
Fax (0925) 828196

(CIS) **Applied Imaging International Ltd,**
Hylton Park, Wessington Way, Sunderland, Tyne and Wear SR5 3HD, UK
Tel. (091) 516 0505
Fax (091) 516 0512

US Address: Applied Imaging Inc., 2340A Walsh Avenue, Santa Clara, CA 95051, USA. Tel. (408) 562 0250; fax (408) 562 0264

(MG) **Bio-Rad Laboratories,**
Bio-Rad House, Maylands Avenue, Hemel Hempstead, Herts HP2 7TD, UK
Tel. (0442) 232552, (0800) 181134
Fax (0442) 259118

(MG, TCR) **Boehringer Mannheim,**
Bell Lane, Lewes, East Sussex BN7 1LG, UK
Tel. (0273) 480444, (0800) 521578
Fax (0273) 480266

US Address: 9115 Hague Road, PO Box 50414, Indianapolis, IN 46250–0414, USA. Tel. (800) 262 1640; fax (317) 576 2754

(CIS) **BRSL (Brian Reece Scientific Ltd),**
12 West Mills, Newbury, Berks RG14 5HG, UK
Tel. (0635) 32827
Fax (0635) 34542

BDH Laboratories, see Merck Ltd

(TCP) **Becton Dickinson Ltd,**
Between Towns Road, Cowley, Oxford OX4 3LY, UK
Tel. (0865) 748844
Fax (0865) 717313

US Address: 2 Bridgewater Lane, Lincoln Park, NJ 07035, USA (Clay Adams Division, 299 Webro Road, Parsippany, NJ 07054, USA. Tel. (201) 887 4800); fax (201) 887 2695

(TCP) **Bibby Sterilin Ltd,**
Tilling Drive, Stone, Staffs ST15 0SA, UK
Tel. (0785) 812121
Fax (0785) 813748

(TCR) **Biological Industries Ltd,**
58 Telford Road, Cumbernauld, Glasgow G67 2AX, UK
Tel. (0236) 728700
Fax (0236) 738001

(MG) **Cambio,**
34 Millington Road, Cambridge, Cambs CB3 9HP, UK
Tel. (0223) 66500
Fax (0223) 350069

(TCP) **Ciba Corning Diagnostics Ltd,**
Colchester Road, Halstead, Essex CO9 2DX, UK
Tel. (0787) 472461
Fax (0787) 475088

(P) **Chromagene Ltd,**
13 Abbey Road, Abbey Mills, Kirkstall, Leeds L55 3HP, UK (or Houldsworth St, Reddish, Stockport SK5 6DA, UK)

Clay Adams, see Becton Dickinson

(TCP) **Costar Europe Ltd,**
Victoria House, 28–38 Desborough Street, High Wycombe, Bucks HP11 2NF, UK
Tel. (0494) 471207
Fax (0494) 459540

(TCR) Difco Laboratories Ltd,
PO Box 14B, Central Avenue, East Molesey, Surrey KT8 0SE, UK

US Address: Difco Laboratories, PO Box 1058, Detroit, MI 48232, USA. Tel. (313) 462 8500; fax (313) 462 8517

(P) Eastman Kodak Ltd,
PO Box 66, Station Road, Hemel Hempstead, Herts HP1 1JU, UK
Tel. (0442) 61122
Fax (0442) 844035

US Address: Eastman Kodak Inc., 25 Science Park, New Haven, CT 06511, USA. Tel. (203) 786 5600, (800) 225 5352; fax (203) 624 3143, (800) 879 4979

(C,E,TCP,TCR) Fisons Scientific Equipment,
Bishop Meadow Road, Loughborough, Leics LE11 0RG, UK
Tel. (0509) 231166
Fax (0509) 231893

(MG) Hybaid Ltd,
111–113 Waldegrave Road, Teddington, Middx TW11 8LL, UK
Tel. (081) 977 3266
Fax (081) 977 0170

(E,TCP,TCR,MG) ICN Flow Biomedicals,
Unit 18, Thame Park Business Centre, Wenman Road, Thame, Oxon OX9 3XA, UK
Tel. (0844) 213366
Fax (0844) 217722

US Address: ICN Biomedicals Inc., 3300 Hyland Avenue, Costa Mesa, CA 92626, USA. Tel. (800) 545 0530; fax (800) 334 6999

(E) IEC Ltd (International Equipment Company),
Unit 7, Lawrence Way, Brewers Hill Road, Dunstable, Beds LU6 1BD, UK
Tel. (0582) 604669
Fax (0582) 609257

US Address: 300 Second Avenue, Needham Heights, MA 02194, USA

(TCR,TCP,MG) **Gibco-BRL Life Technologies,**
Unit 4, Cowley Trading Estate, Longbridge Way, Uxbridge, Middx UB8 2YG, UK

US Address: 2000 Alfred Nobel Drive, Hercules, CA 94547, USA. Tel. (516) 756 2575; fax (516) 756 2594

(R,TCP) **Heraeus Equipment Ltd,**
9 Wates Way, Ongar Way, Brentwood, Essex CM15 9TB, UK
Tel. (0277) 231511
Fax (0277) 261856

US Address: Heraeus Instruments Inc., 111A Corporate Blvd, South Plainfield, NJ 07080, USA. Tel. (908) 754 0100; fax (908) 754 9494

(E,TCP) **A.R. Horwell Ltd,**
73 Maygrove Road, London NW6 2BP, UK
Tel. (071) 328 1551
Fax (071) 372 5259

(CIS) **Imagenetics,**
31 New York Avenue, Framingham, MA 01701-9883, USA
Tel. (508) 872 3113
Fax (508) 872 3420

(TCR) **Imperial Laboratories (Europe) Ltd,**
West Portway, Andover, Hants SP10 3LF, UK
Tel. (0264) 333311
Fax (0264) 332412

(E) **Jencons (Scientific) Ltd,**
Cherrycourt Way Industrial Estate, Stanbridge Road, Leighton Buzzard LU7 8UA, UK
Tel. (0525) 372010
Fax (0525) 379547

(E) **Leec Ltd,**
Private Road No. 7, Colwick Industrial Estate, Nottingham NG4 2AJ, UK
Tel. (0602) 616222
Fax (0602) 616680

(MIC) **Leica Cambridge Ltd,**
Clifton Road, Cambridge CB1 3QH, UK
Tel. (0223) 411411
Fax (0223) 412776

Life Technologies Ltd, see Gibco-BRL Life Technologies

(E) **LKB Instruments Ltd,**
LKB House, 232 Addington Road, Selsdon, South Croydon, Surrey CR2 8YD, UK

US Address: 9319 Gaither Road, Gaithersburgh, MD 20877, USA

(E) **MDH Ltd (Microflow, Dent and Hellyer),**
Walworth Road, Andover, Hants SP10 5AA, UK
Tel. (0264) 362111
Fax (0264) 536452

(E,TCP) **Millipore Corporation,**
The Boulevard, Blackmore Lane, Watford, Herts WD1 8YW, UK
Tel. (0923) 816375
Fax (0923) 818297

US Address: PO Box 255, Bedford, MA 01730, USA. Tel. (617) 275 9200, (800) 225 1380; fax (617) 271 0290

(MIC,CIS) **Nikon (UK) Ltd,**
Instruments Division, Nikon House, 380 Richmond Road, Kingston, Surrey KT2 5PR, UK

US Address: Nikon Inc. Instrument Group, 1300 Walt Whitman Road, Melville, NY 11747–3064, USA. Tel. (516) 547 8500; fax (516) 547 0306

(TCR) **Northumbria Biologicals Ltd,**
Nelson Industrial Estate, Cramlington, Northumberland NE23 9BL, UK
Tel. (0670) 732992
Fax (0670) 732537

(E) **Medical Air Technology Ltd,**
Canto House, Wilton Street, Denton, Manchester M34 3LZ, UK
Tel. (061) 320 5652
Fax (061) 335 0313

(C,TCR) **Merck Ltd,**
Merck House, Seldown Lane, Poole, Dorset BH15 1TD, UK
Tel. (0202) 669700, (0800) 223344
Fax (0202) 665599

(TCR,MG) **Metachem Diagnostics Ltd,**
29 Forest Road, Piddington, Northampton NN7 2DA, UK
Tel. (0604) 870370
Fax (0604) 870194

Nunc, see Gibco-BRL Life Technologies

(MIC) **Olympus Optical Co. (UK) Ltd,**
Biomedical Products Division, 7 West Links, Tollgate, Eastleigh, Hants SO5 3TG, UK
Tel. (0703) 644199
Fax (0703) 612908

US Address: Olympus America Inc., Clinical Instruments Division, 4 Nevada Drive, Lake Success, NY 11042–1179, USA. Tel. (516) 488 3880, (800) 446 5967; Fax (516) 222 7920

Oncor Instrument Systems,
9581 Ridgehaven Court, San Diego, CA 92123, USA
Tel. (619) 560 9355
Fax (619) 560 9584

UK Supplier: see Alpha Laboratories Ltd

(CIS) **Optivision (Yorkshire) Ltd,**
Ahed House, Dewsbury Road, Ossett, W. Yorks WF5 9ND, UK
Tel. (0924) 277727
Fax (0924) 280016

(CIS) **Perceptive Scientific International Ltd,**
Halladale, Lakeside, Chester Business Park, Wrexham Road, Chester CH4 9QT, UK
Tel. (0244) 682288
Fax (0244) 681555

US Address: 2525 South Shore Blvd, Ste. 100, League City, TX 77573, USA. Tel. (713) 334 3027; fax (713) 334 3116

(TCR) **Sera-lab Ltd,**
Hophurst Lane, Crawley Down, Sussex RH10 4FF, UK
Tel. (0342) 716366
Fax (0342) 717351

(E) **Technicon Instruments Ltd,**
Evans House, Hamilton Close, Basingstoke, Hants RG21 2YE, UK
Tel. (0256) 29181
Fax (0256) 52916

(MG) **Vector Laboratories Ltd,**
16 Wulfric Square, Bretton, Peterborough PE3 8BR, UK
Tel. (0733) 265530
Fax (0733) 263048

(TCR) **Wellcome Diagnostics,**
Temple Hill, Dartford, Kent DA1H 5AH, UK

(E) **Whatman International Ltd,**
Whatman House, St Leonards Road, 20/20 Maidstone, Kent ME16 OLS, UK
Tel. (0622) 676670
Fax (0622) 677011

(C,TCR) Sigma Chemical Company Ltd,
Fancy Road, Poole, Dorset BH17 7NH, UK
Tel. (0202) 733114, (0800) 447788. Overseas call collect (0202) 733114
Fax (0202) 715460

US Address: Sigma Chemical Company, PO Box 14508, 3500 DeKalb Street, St Louis, MO 63178, USA. Tel. (800) 848 7791, overseas call reverse charge (314) 771 5765

Sterilin, see Bibby Sterilin Ltd

(CIS) Tecan UK Ltd,
18 The High Street, Goring-on-Thames, Reading, Berks RG8 9AR, UK
Tel. (0491) 875087
Fax (0491) 875432

(E) Don Whitley Scientific Ltd,
14 Otley Road, Shipley, W. Yorks BD17 7SE, UK
Tel. (0274) 595728
Fax (0274) 531197

(MIC) Carl Zeiss Scientific Instruments,
Zeiss England House, 1 Elstree Way, Borehamwood, Herts WD6 1NH, UK
Tel. (081) 953 1688
Fax (081) 953 9456

US Address: Carl Zeiss Inc., Microscope Division, 1 Zeiss Drive, Thornwood, NY 10594, USA. Tel. (800) 233 2343; fax (914) 681 7446

REFERENCES

Chapter 1

1. ISCN (1985) in *An International System for Human Cytogenetic Nomenclature* (D.G. Harnden and H.P. Klinger, eds). Karger, Basel.
2. Sutherland, G.R. and Ledbetter, D.H. (1989) *Cytogenet. Cell Genet.* **51**, 452.
3. Warburton, D., Byrne, J. and Canki, N. (1991) *Chromosome Anomalies and Prenatal Development: an Atlas.* Oxford University Press, Oxford.
4. Ledbetter, D.H. *et al.* (1992) *Prenat. Diagn.* **12**, 317.
5. ACC Working Party on Chorionic Villi in Prenatal Diagnosis (1994) *Prenat. Diagn.* **14**, 363.
6. Warburton, D., Kline, J., Stein, Z. and Strombino, B. (1986) in *Perinatal Genetics: Diagnosis and Treatment* (I.H. Porter, W.H. Hatcher and A.M. Willey, eds). Birth Defects Institute Symposia. Academic Press, Orlando, FL.
7. Ferguson-Smith, M.A. and Yates, J.R.W. (1984) *Prenat.*

18. Fryns, J.P., Kleczkowska, A., Lagae, L., Kenis, H. and Van Den Berge, H. (1993) *Ann. Génét.* **36**, 129.
19. Lindor, N.M., Jalal, S.M., Thibodeau, S.N., Bonde, D., Sauser, K.L. and Karnes, P.S. (1993) *Clin. Genet.* **44**, 185.
20. Djalali, M., Barbi, G. and Grab, D. (1991) *Prenat. Diagn.* **11**, 399.
21. Rethoré, M.-O., Debray, P., Guesne, M.-C., Amédéc-Manesme, O., Iris, L. and Lejeune, J. (1981) *Ann. Génét.* **24**, 34.
22. Hsu, L.Y.F., Kaffe, S. and Perlis, T.E. (1991) *Prenat. Diagn.* **11**, 7.
23. Kobrynski, L., *et al.* (1993) *Am. J. Med. Genet.* **46**, 68.
24. Wolstenholme, J., Crocker, M. and Jonasson, J. (1988) *Prenat. Diagn.* **8**, 339.
25. Palmer, C.G. and Reichmann, A. (1976) *Hum. Genet.* **35**, 35.
26. Hassold, T.J., Jacobs, P.A., Leppert, M. and Sheldon, M. (1987) *J. Med. Genet.* **24**, 725.
27. Hassold, T.J., Pettay, D., Freeman, S.B., Grantham, M. and Takaesu, N. (1991) *J. Med. Genet.* **28**, 159.
28. Nöthen, M.M., Eggermann, T., Erdmann, J., Eiben, B.,

Diagn. **4** (special issue), 5.
8. Hsu, L.Y.F. and Perlis, T.E. (1984) *Prenat. Diagn.* **4** (special issue), 97.
9. Casey, J., Ketterer, D.M., Heisler, K.L., Daugherty, E.A., Prince, P.M. and Giles, H.R. (1990) *Am. J. Hum. Genet.* **47** (Suppl.), A270.
10. De Keyser, F., Matthys, E., De Paepe, A., Verschraegen-Spae, M.R. and Matton, M. (1988) *J. Med. Genet.* **25**, 358.
11. Reddy, K.S., Blakemore, K.J., Stetten, G. and Corson, V. (1990) *Prenat. Diagn.* **10**, 417.
12. Tarani, L., Colloridi, F., Raguso, G., Rizzuti, A., Bruni, L., Tozzi, M.C., Palermo, D., Panero, A. and Vignetti, P. (1994) *Ann. Genet.* **37**, 14.
13. Mantagos, S., McReynolds, J.W., Seashore, M.R. and Breg, W.R. (1981) *J. Med. Genet.* **18**, 377.
14. de France, H.F., Beemer, F.A., Senders, R.Ch. and Schaminée-Main, S.C.E. (1985) *Clin. Genet.* **27**, 92.
15. English, C.J., Goodship, J.A., Jackson, A., Lowry, M. and Wolstenholme, J. (1994) *J. Med. Genet.* **31**, 253.
16. Vachvanichsanong, P., Jinorose, U. and Sangnuachua, P. (1991) *Am. J. Med. Genet.* **40**, 80.
17. Coldwell, S., Fitzgerald, B., Semmens, J.M., Ede, R. and Bateman, C. (1981) *J. Med. Genet.* **18**, 146.

   Hofmann, D., Propping, P. and Schwanitz, G. (1993) *Hum. Genet.* **92**, 347.
29. Fisher, J.M., Harvey, J.F., Lindenbaum, R.H., Boyd, P.A. and Jacobs, P.A. (1993) *Am. J. Hum. Genet.* **52**, 1139.
30. Antonarakis, S.E. et al. (1992) *Am. J. Hum. Genet.* **50**, 544.
31. Mathur, A., Stekol, L., Schatz, D., MacLaren, N.K., Scott, M.L. and Lippe, B. (1991) *Am. J. Hum. Genet.* **48**, 682.
32. Lorda-Sanchez, I., Binkert, F., Maechler, M. and Schinzel, A. (1991) *Am. J. Hum. Genet.* **49**, 1034.
33. May, K.M. et al. (1990) *Am. J. Hum. Genet.* **46**, 754.
34. Hassold, T.J., Sherman, S.L., Pettay, D., Page, D.C. and Jacobs, P.A. (1991) *Am. J. Hum. Genet.* **49**, 253.
35. Jacobs, P.A., Szulman, A.E., Funkhouser, J., Matsuura, J. and Wilson, C.C. (1982) *Ann. Hum. Genet.* **46**, 223.
36. McFadden, D.E., Kwong, L.C., Yam, I.Y.L. and Langois, S. (1993) *Hum. Genet.* **92**, 465.
37. Angell, R.R., Xian, J., Keith, J., Ledger, W. and Baird, D.T. (1994) *Cytogenet. Cell Genet.* **65**, 194.
38. Buckton, K.E., O'Riordan, M.L., Ratcliffe, S., Slight, J., Mitchell, M., McBeath, S., Keay, A.J., Barr, D. and Short, M. (1980) *Ann. Hum. Genet.* **43**, 227.
39. Hansteen, I.-L., Varslot, K., Steen-Johnsen, J. and Langard, S. (1982) *Clin. Genet.* **21**, 309.

40. Nielsen, J., Wohlert, M., Faaborg-Andersen, J., Hansen, K.B., Hvidman, L., Krag-Olsen, B., Moulvad, I. and Videbech, P. (1982) *Hum. Genet.* **61**, 98.
41. Tawn, E.J. and Earl, R. (1992) *Mutat. Res.* **283**, 69.
42. Hook, E.B., Healy, N.P. and Willey, A.M. (1989) *Ann. Hum. Genet.* **53**, 237.
43. Madan, K. and Menko, F.H. (1992) *Hum. Genet.* **89**, 1.
44. Kosztolanyi, G., Mehes, K. and Hook, E.B. (1991) *Hum. Genet.* **87**, 320.
45. Hook, E.B. (1993) in *Prenatal Diagnosis and Screening* (D.J.H. Brock, C.H. Rodeck and M.A. Ferguson-Smith, eds), p. 351. Churchill Livingstone, Edinburgh.
46. Boué, A. and Gallano, P. (1984) *Prenat. Diagn.* **4** (special issue), 45.
47. Unpublished data. Report of the Association of Clinical Cytogeneticists UK Working Party on Chorionic Villi in Prenatal Diagnosis 1987–89.
48. Iselius, L. *et al.* (1983) *Hum. Genet.* **64**, 343.

Chapter 2

1. Bennett, J.M., Catovsky, D., Daniel, M.T., Flandrin, G.,

International Histological Classification of Tumours, No. 14. World Health Organization, Geneva.
13. Rappaport, H. (1986) in *Atlas of Tumor Pathology*, p. 270. Armed Forces Institute of Pathology, Washington, DC.

Chapter 3

1. Fisher, F. and Scambler, P. (1994) *Hum. Genet.* **7,** 5.
2. Tommerup, N. (1993) *J. Med. Genet.* **30**, 713.
3. McKusick, V.A. and Amberger, J.S. (1993) *J. Med. Genet.* **30**, 1.
4. Chen, Q. *et al.* (1990) *EMBO J.* **9**, 415.
5. Kamps, M.P. *et al.* (1990) *Cell* **60**, 547.
6. Bocian, M. and Walker, A.P. (1987) *Am. J. Med. Genet.* **26**, 437.
7. Hecht, B.K.M. *et al.* (1991) *Am. J. Med. Genet.* **40**, 130.
8. Phipps, M.E. (1994) *et al. Hum. Mol. Genet.* **3**, 903.
9. Gemmill, M. *et al.* (1993) *Am. J. Hum. Genet.* **53**, 24.
10. Altherr, M.H. *et al.* (1991) *Am. J. Hum. Genet.* **49**, 1235.
11. Altherr, M.R. *et al.* (1992) *Am. J. Med. Genet.* **44**, 449.
12. Lacassie, Y. *et al.* (1977) *Am. J. Hum. Genet.* **29**, 641.
13. Griesinger, F. *et al.* (1994) *Leukemia* **8,** 542.
14. Overhauser, J. *et al.* (1994) *Hum. Mol. Genet.* **3**, 247.
15. Lawson, S.V. *et al.* (1991) *Genes, Chrom. Cancer* **3**, 382.

Galton, D.A.G., Gralnik, H.R. and Sultan, C. (1976) *Br. J. Haemat.* **33**, 451.
2. Second MIC Cooperative Study Group (1988) *Br. J. Haemat.* **68**, 487.
3. Bennett, J.M., Catovsky, D., Daniel, M.T., Flandrin, G., Galton, D.A.G., Gralnik, H.R. and Sultan, C. (1985) *Ann. Intern. Med.* **103**, 620.
4. Bain, B.J. (1990) *Leukaemia Diagnosis: a Guide to the FAB Classification*. Gower Medical Publishing, London.
5. Bennett, J.M., Catovsky, D., Daniel, M.T., Flandrin, G., Galton, D.A.G., Gralnik, H.R. and Sultan, C. (1982) *Br. J. Haemat.* **51**, 189.
6. Third MIC Cooperative Study Group (1988) *Cancer Genet. Cytogenet.* **32**, 1.
7. First MIC Cooperative Study Group (1986) *Cancer Genet. Cytogenet.* **23**, 189.
8. Bennett, M.H., Farrer-Brown, G., Henry, K. and Jelliffe, A.M. (1974) *Lancet* **ii**, 405.
9. Dorfman, R.F. (1974) *Lancet* **ii**, 961.
10. Lennert, K., Mohri, N. Stein, H. and Kaiseling, E. (1975) *Br. J. Haemat.* **31** (Suppl.) 193.
11. Lukes, R.J. and Collins, R.D. (1974) *Cancer* **34**, 1488.
12. Mathe, G., Rappaport, H. and O'Connor, G.T. (1976) in *WHO*
16. Ward, J.R.T. *et al.* (1993) *Genomics* **17**, 15.
17. Lindgren, V. *et al.* (1992) *Am. J. Hum. Genet.* **50**, 988.
18. von Lindern, M. *et al.* (1992) *Genes, Chrom. Cancer* **5**, 227.
19. Reardon, R. and Winter, R. (1994) *J. Med. Genet.* **31**, 393.
20. Vorkamp, A. *et al.* (1991) *Nature* **352**, 539.
21. Brueton, L. *et al.* (1988) *Am. J. Med. Genet.* **31**, 799.
22. Pettigrew, A.L. *et al.* (1991) *Hum. Genet.* **87**, 452.
23. Wagner, K. *et al.* (1990) *Genomics* **8**, 487.
24. Brueton, L.A. *et al.* (1992) *J. Med. Genet.* **29**, 681.
25. Curran, M.E. *et al.* (1993) *Cell* **73**, 159.
26. Ewart, A.E. *et al.* (1993) *Hum. Mol. Genet.* **5**, 11.
27. Naritomi, K. *et al.* (1989) *Hum. Genet.* **84**, 79.
28. Gurrieri, F. *et al.* (1993) *Nature Genetics* **3**, 247.
29. Lux, S.E. *et al.* (1990) *Nature* **345**, 736.
30. Erickson, P. *et al.* (1992) *Blood* **80,** 1825.
31. Fryns, J.P. *et al.* (1986) *Hum. Genet.* **74**,188.
32. Ludecke, H.J. *et al.* (1991) *Am. J. Hum. Genet.* **49**, 1197.
33. Ludecke, H.J. *et al.* (1989) *Hum. Genet.* **82**, 327.
34. Verlander, P.C. *et al.* (1993) *Am. J. Hum. Genet.* **53**, 11.
35. Trachuk, D.C. *et al.* (1990) *Science* **250**, 559.
36. Cross, N.C.P. *et al.* (1994) *Leukemia* **8**, 186.
37. Monaco, A.P. *et al.* (1986) *Nature* **323**, 646.
38. Huff, V. *et al.* (1990) *Hum. Genet.* **84**, 253.

39. van Heyningen, V. and Hastie, N. (1992) *Trends Genet.* **8**, 16.
40. Gessler, M. *et al.* (1989) *Am. J. Hum. Genet.* **44**, 486.
41. Breuning, W. *et al.* (1992) *Nature Genetics* **1**, 144.
42. Fantes, J.A. *et al.* (1992) *Am. J. Hum. Genet.* **51**, 1286.
43. Slee, J.J. *et al.* (1991) *J. Med. Genet.* **28**, 413.
44. Horsthemke, B. (1992) *Cancer Genet. Cytogenet.* **63**, 1.
45. Kloss, K. *et al.* (1991) *Am. J. Med. Genet.* **39**, 196.
46. Goddard, A.D. *et al.* (1990) *Clin. Genet.* **37**, 117.
47. Wiggs, J. *et al.* (1988) *New Engl. J. Med.* **318**, 151.
48. Nicholls, R.D. (1993) *Am. J. Med. Genet.* **46**, 16.
49. Hamabe, J. *et al.* (1991) *Am. J. Med. Genet.* **41**, 54.
50. Mutirangura, A. *et al.* (1993) *Hum. Mol. Genet.* **2**, 143.
51. Dittrich, B. *et al.* (1992) *Hum. Genet.* **90**, 313.
52. de The, H. *et al.* (1991) *Cell* **66**, 675.
53. Gibbons, R.J. *et al.* (1991) *J. Med. Genet.* **28**, 729.
54. Patel, P.I. and Lupski, J.R. (1994) *Trends Genet.* **10**, 128.
55. Greenberg, F. *et al.* (1991) *Am. J. Hum. Genet.* **49**, 1207.
56. Reiner, O. *et al.* (1993) *Nature* **364**, 717.
57. Kuwano, A. *et al.* (1991) *Am. J. Hum. Genet.* **49**, 707.
58. Dobyns, W.B. *et al.* (1991) *Am. J. Hum. Genet.* **48**, 584.
59. Ledbetter, S.A. *et al.* (1992) *Am. J. Hum. Genet.* **50**, 182.
60. Lupski, J.R. *et al.* (1991) *Cell* **66**, 219.
61. Valentijn, L.J. (1992) *Nature Genetics*, **2**, 288.

85. Pillers, D.A.M. *et al.* (1990) *Am. J. Med. Genet.* **36**, 23.
86. Rousseau, F. *et al.* (1991) *New Engl. J. Med.* **325**, 1673.
87. Mulley, J.C. *et al.* (1992) *J. Med. Genet.* **29**, 368.
88. Sutherland, G.R. and Baker, E. (1992) *Hum. Mol. Genet.* **1**, 111.
89. Hirst, M.C. *et al.* (1993) *Hum. Mol. Genet.* **2**, 197.
90. Hawkins, J.R. *et al.* (1992) *Am. J. Hum. Genet.* **51**, 979.
91. Pereira, E.T. *et al.* (1991) *J. Med. Genet.* **28**, 591.

Chapter 4

1. Verma, R.S. and Babu, A. (1989) *Human Chromosomes: Manual of Basic Techniques*, Chapters 3 and 4. Pergamon Press, New York.
2. Rooney, D.E. and Czepulkowski, B.H. (1992) *Human Cytogenetics: a Practical Approach*, Chapters 4 and 8. IRL Press, Oxford.

Chapter 5

1. Gosden, C. (1993) in *Prenatal Diagnosis and Screening* (D.J.H. Brock, C.H. Rodeck and M.A. Ferguson-Smith, eds). Churchill Livingstone, Edinburgh.

62. Wallace, M.R. *et al.* (1990) *Science* **249**, 81.
63. Hong Shen, M. *et al.* (1993) *Hum. Mol. Genet.* **2**, 1861.
64. Rodenhiser, D.I. *et al.* (1993) *J. Med. Genet.* **30**, 363.
65. Viskochil, D. *et al.* (1990) *Cell* **62**, 187.
66. Devaraj, P.E. *et al.* (1994) *Leukemia* **8**, 1131.
67. Anad, F. *et al.* (1990) *J. Med. Genet.* **27**, 729.
68. Schnittger, S. *et al.* (1989) *Hum. Genet.* **83**, 239.
69. Dhorne-Pollet, S. *et al.* (1994) *J. Med. Genet.* **31**, 453.
70. Petersen, M.B. *et al.* (1992) *Am. J. Hum. Genet.* **51**, 516.
71. Pangalos, C.G. *et al.* (1992) *Am. J. Hum. Genet.* **51**, 1015.
72. Pertl, B. *et al.* (1994) *Lancet* **343**, 1197.
73. Scambler, P.J. (1993) *Ann. Cardiac Surg.* 542.
74. Cross, I. *et al.* (1992) *Am. J. Hum. Genet.* **51**, 957.
75. Carey, A.H. *et al.* (1990) *Genomics* **7**, 299.
76. Driscoll, D.A. *et al.* (1992) *Am. J. Hum. Genet.* **50**, 924.
77. Scambler, P.J. *et al.* (1991) *Genomics* **10**, 201.
78. Wirth, B. *et al.* (1988) *Hum. Genet.* **80**, 191.
79. Franco, B. *et al.* (1991) *Nature* **353**, 529.
80. Hardelin, J.P. *et al.* (1993) *Hum. Mol. Genet.* **2**, 373.
81. de Saint-Basile, G. *et al.* (1988) *Hum. Genet.* **80,** P85.
82. Boyd, Y. *et al.* (1988) *Cytogenet. Cell Genet.* **48**, 28.
83. Chamberlain, J.S. *et al.* (1988) *Nucleic Acids Res.* **16**, 11141.
84. Mettinger, T. *et al.* (1988) *Genomics* **3**, 315.

2. Rooney, D.E. and Czepulkowski, B.H. (1992) *Human Cytogenetics: a Practical Approach*. IRL Press, Oxford.
3. Verma, R.S. and Babu, A. (1989) *Human Chromosomes: Manual of Basic Techniques*. Pergamon Press, New York.

Chapter 6

1. Simoni, G., Brambati, B., Danesino, C., Rossella, F., Terzoli, G.L., Ferrari, M. and Fraccaro, M. (1983) *Hum. Genet.* **63**, 349.
2. Heaton, D.E., Czepulkowski, B.H., Horwell, D.H. and Coleman, D.V. (1984) *Prenat. Diagn.* **4**, 279.
3. Smidt-Jensen, S., Christensen, B. and Lind, A.-M. (1989) *Prenat. Diagn.* **9**, 349.
4. Kalousek, D.K., Dill, F.J., Pantzar, T., McGillivray, B.C., Yong, S.L. and Wilson, R.D. (1987) *Hum. Genet.* **77**, 163.
5. Crane, J.P. and Cheung, S.W. (1988) *Prenat. Diagn.* **8**, 119.
6. Vejerslev, L.O. and Mikkelsen, M. (1989) *Prenat. Diagn.* **9**, 575.
7. Ledbetter, D.H. *et al.* (1992) *Prenat. Diagn.* **12**, 317.
8. Hsu, L.Y. and Perlis, T.E. (1984) *Prenat. Diagn.* **4**, 97.
9. Worton, R.G. and Stern, R. (1984) *Prenat. Diagn.* **4**, 131.
10. Bui, T.-H., Iselius, L. and Lindsten, J. (1984) *Prenat. Diagn.* **4**, 145.

11. Claussen, U., Schafer, H. and Trampisch, H.J. (1984) *Hum. Genet.* **67**, 23.
12. Cheung, S.W., Spitznagel, E., Featherstone, T. and Crane, J.P. (1990) *Prenat. Diagn.* **10**, 41.
13. Hook, E.B. (1981) *Obstet. Gynaecol.* **58**, 282.
14. Gardner, R.J.M. and Sutherland, G.R. (1989) *Chromosome Abnormalities and Genetic Counselling*. Oxford University Press, New York.
15. Cuckle, H.S. and Wald, N.J. (1990) in *Prenatal Diagnosis and Prognosis* (R. Lilford, ed.), p. 67. Butterworth, London.
16. Wald, N.J. and Cuckle, H.S. (1993) in *Prenatal Diagnosis and Screening* (D.J.H. Brock, C.H. Rodeck and M.A. Ferguson-Smith, eds), p. 563. Churchill Livingstone, Edinburgh.
17. Gosden, C. (1990) in *Prenatal Diagnosis and Prognosis* (R. Lilford, ed.), p. 104. Butterworth, London.
18. Teshima, I.E. *et al.* (1992) *Prenat. Diagn.* **12**, 443.
19. Kalousek, D.K. (1988) *Growth, Genet. Hormones* **4**, 1.
20. Association of Clinical Cytogeneticists Working Party on Chorionic Villi in Prenatal Diagnosis (1994) *Prenat. Diagn.* **14**, 363.
21. Hook, E.B. (1977) *Am. J. Hum. Genet.* **29**, 94.
22. Hook, E.B. (1993) in *Prenatal Diagnosis and Screening* (D.J.H. Brock, C.H. Rodeck and M.A. Ferguson-Smith, eds), p 351. Churchill Livingstone, Edinburgh.

8. HMSO (1989) *Guidelines for the Testing of Chemicals for Mutagenicity*. Department of Health Report on Health and Social Subjects, no. 35. HMSO, London.
9. Kirkland, D.J. (ed.) (1990) *Basic Mutagenicity Tests: UKEMS Recommended Procedures*. Cambridge University Press, Cambridge.
10. Howell, R.T. and Taylor, A.M.R. (1992) in *Human Cytogenetics: a Practical Approach*, Vol. II (D.E. Rooney and B.H. Czepulkowski, eds), p. 209. IRL Press, Oxford.
11. McKinnon, P.J. (1987) *Hum. Genet.* **75**, 197.
12. Woods, C.G., Bundey, S.E. and Taylor, A.M.R. (1990) *Hum. Genet.* **84**, 555.
13. Passarge, E. (1983) in *Chromosome Mutation and Neoplasia* (J. German, ed.), p. 11. Alan R. Liss, New York.
14. German, J. and Passarge, E. (1989) *Clin. Genet.* **35**, 57.
15. Schroeder-Kurth, T.M., Auerback, A.D. and Obe, G. (eds) (1989) *Fanconi Anaemia. Clinical, Cytogenetic and Experimental Aspects*. Springer Verlag, Berlin.
16. Taalman, R.D.F.M., Hustinx, T.W.J., Weemaes, C.M.R., Seemanova, E., Schmidt, A., Passarge, E. and Scheres, J.M.J.C. (1989) *Am. J. Med. Genet.* **32**, 425.
17. Maraschio, P., Zuffardi, U., Dalla Fior. T. and Ticpolo, L. (1988) *J. Med. Genet.* **25**, 173.

23. Cunningham, F.G., MacDonald, P.C., Leveno, K.J., Gant, N.F. and Gilstrap, L.C. (eds) (1993) *Williams Obstetrics*. Prentice-Hall, London.

Chapter 7

1. Tawn, E.J. and Holdsworth, D. (1992) in *Human Cytogenetics: a Practical Approach*, Vol. II (D.E. Rooney and B.H. Czepulkowski, eds), p. 189. IRL Press, Oxford.
2. Savage, J.R.K. (1975) *J. Med. Genet.* **13**, 103.
3. Lloyd, D.C. and Tawn, E.J. (1989) *Clin. Cytogen. Bull.* **2** (3), 54.
4. Straume, T., Lucas, J.N., Tucker, J.D. and Bigbee, W.L. (1992) *Health Physics* **62**, 122.
5. WHO (1985) *Guidelines for the Study of Genetic Effects in Human Populations*. Environmental Health Criteria, 46, World Health Organization, Geneva.
6. Carrano, A.V. and Natarajan, A.T. (1988) *Mutat. Res.* **204**, 379.
7. Venitt, S. and Parry, J.M. (eds) (1984) *Mutagenicity Testing: a Practical Approach*. IRL Press, Oxford.

18. Van Den Berg, D.J. and Francke, U. (1993) *Am. J. Med. Genet.* **47**, 1104.
19. Gebhart, E., Bauer, R., Raub, U., Schinzel, M., Ruprecht, K.W. and Jones, J.B. (1988) *Hum. Genet.* **80**, 135.
20. Cleaver, J.E. and Kraemer, K.H. (1989) in *The Metabolic Basis of Inherited Disease*, Vol. II (C.R. Scriber, A.L. Beaudet, W.S. Sly and D. Valle, eds), p. 2949. McGraw-Hill, New York.
21. Nance, M.A. and Berry, S.A. (1992) *Am. J. Med. Genet.* **42**, 68.
22. Lehman, A.R. (1987) *Cancer Rev.* **7**, 82.

Chapter 8

1. HMSO (1991) *Safe Working and the Prevention of Infection in Clinical Laboratories*, Section 18.2. HMSO, London.
2. HMSO (1991) *Safe Working and the Prevention of Infection in Clinical Laboratories*, Section 5.2. HMSO, London.
3. HMSO (1988) *Control of Substances Hazardous to Health (general ACoP)*, Section 5.3. HMSO, London.
4. HMSO (1993) *The Chemicals (Hazard Information and Packaging) Regulations*, Section 18.3. HMSO, London.

FURTHER READING

Chapter 2

Bain, B. (1990) *Leukaemia Diagnosis*. Gower Medical Publishing, London.

Goldman, J.M. and Preisler, H.D. (eds) (1984) *Leukaemia Haematology*. Butterworths International Medical Reviews, London.

Heim, S. and Mitelman, F. (1987) *Cancer Cytogenetics*. Alan R. Liss, New York.

Rooney, D.E. and Czepulkowski, B.H. (eds) (1992) *Human Cytogenetics: a Practical Approach*, Vol. II. IRL Press, Oxford.

142   *Human Cytogenetics*

APPENDIX

Ideogram of G-banded human karyotype

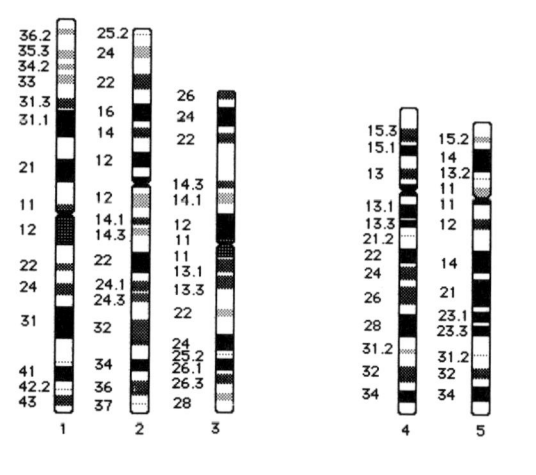

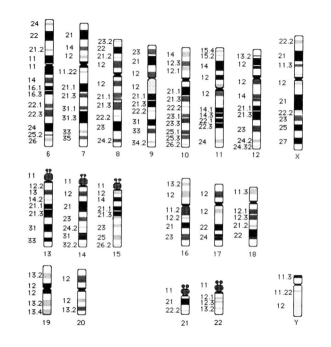

INDEX

Aberration induction, 116
Aberrations – *see* Chromatid-type *and*
    Chromosome-type
Absorption spectra
  of fluorochromes, 61, 75
Acentric, 16, 102
Acetic acid, 120, 121
Achromatic lens – *see* Lens
Acridine orange, 121
Acrocentric chromosome – *see*
    Chromosome
Actinomycin D, 121
Acute leukemia
  biphenotypic, 25
  symptoms of, 34
Acute lymphoblastic leukemia (ALL),
    24–26, 77, 86, 87, 88, 109

prostatic, 33
Adriamycin (doxorubicin), 121
Adult T-cell leukemia/lymphoma, 28,
    86
Ageing, premature, 109
Alagille syndrome, 51
Alkylating agents, 81, 106, 110, 114,
    116
  diepoxybutane, 81, 107, 110, 121
  mitomycin C, 81, 107, 110, 122
  mustine hydrochloride (nitrogen
      mustard), 81, 122
Alleles
  parental, 42, 43, 50
Alphafetoprotein (AFP), 89, 95, 96
Amethopterin – *see* Methotrexate
Amino acids, 78

Angelman syndrome, 48
Aniridia, 47
Annular stop, 57
Anorexia, 34, 37, 39
Antibiotics, 79
  penicillin, 79
  streptomycin, 79
Anticoagulants
  heparin, 79
Anti-condensing agents – *see*
    Actinomycin D, 5′ Azacytidine
    and Ethidium bromide
Antimetabolites, 115, 116
Aperture iris diaphragm
  adjustment, 57
Aphidicolin – *see* Fragile sites
Apochromat lens – *see* Lens

chromosome changes in, 24, 25
FAB classification, 26
in molecular genetics, 42, 43, 50
translocations and oncogenes, 29, 30
Acute lymphocytic leukemia (ALL) – *see* Acute lymphoblastic leukemia
Acute lymphoid leukemia (ALL) – *see* Acute lymphoblastic leukemia
Acute myeloblastic leukemia (AML) – *see* Acute myeloid leukemia
Acute myeloid leukemia (AML), 19–21, 86, 87, 88
chromosome abnormalities in, 20, 21
FAB classification, 19
in molecular genetics, 43, 45, 49
translocations and oncogenes, 29
Acute promyelocytic leukemia (APL), 19–21, 86
Adenocarcinoma, 32
Amniocentesis, 91, 94
incidence of constitutional abnormalities, 7–9
segregation patterns of Robertsonian translocations, 13
Amniotic fluid
cells, 83
culture, 83
cultured cell types, properties and morphology, 83
incidence of structural chromosome abnormalities, 11
Anemia, 22, 34, 35, 36, 38, 39, 88
macrocytic, 37
megaloblastic, 87
normochromic/normocytic, 34, 35, 37
see also Fanconi's anemia
Aneuploidy, 115

ASA rating, 64
Ascites, 97
Askin tumor, 32
Ataxia telangiectasia (AT), 81, 106, 107, 109, 110, 116
Atrial septal defect (ASD) – *see* Cardiovascular defects
Atypical lipoma, 31
Auer rod, 20, 22, 34
5′ Azacytidine, 121
see also Fragile sites

Bacteria
disinfectants active against, 119
spores, disinfectants active against, 119
Banding
C-banding, 68, 81
differential replication staining, 72

distamycin A–DAPI (DA–DAPI), 70
fluorescence plus Giemsa, 102
G-banding, 66
G-11 banding, 69
high-resolution, 82
kinetochore staining, 71
Q-banding, 66
R-banding, 67
restriction endonuclease Giemsa staining, 70
silver staining, 69
T-banding, 70
X chromatin identification, 71
Y chromatin identification, 72
Barr body, 71
Barrier filter – *see* Filter
Base analogs, 114
Basophils, 84

B-lineage
ALL, 24, 26
CLL, 28
lymphoma, 27
Bloom's syndrome, 81, 106, 109, 110
Bone marrow sample
culture, 78
failure, 87
potential diagnoses, 87–88
reasons for referral, 87–88
seeding volumes, 85
setting up culture, 77, 86
Branchio-oto-renal syndrome, 45
Breast carcinoma – *see* Carcinoma
5′-Bromodeoxyuridine (BrdU), 72, 81, 82, 102, 116, 121
Buccal smear, 71, 72

Cardiovascular defects, 97
atrial septal defect, 97
ventricular septal defect, 97
CATCH 22, 51
C-banding – *see* Banding
CD4
and permissive cells, 119
receptor, 112
Cell
cycle
G_0, G_1, 106
G_2, 81, 106
freezing, 79
synchronization, 80, 86, 116, 121, 122
synchronizing agents
5′-fluorodeoxyuridine (FdU), 78, 80, 116, 121
methotrexate (MTX), 78, 80, 121, 122

B-cell
  ALL, 24
  translocation, 29, 30
Beckwith–Wiedemann syndrome, 46
Bence–Jones protein, 37
Benign ovarian tumor – *see* Tumor
Biparietal diameter, 91, 100–101
Biphenotypic acute leukemia, 25
Birth
  incidence of constitutional abnormalities, 7–9
  incidence of structural chromosome abnormalities, 10–11
Bladder carcinoma – *see* Carcinoma
Blast cells, 20, 22, 26, 34
  in blood, potential diagnosis, 87
Bleomycin, 81, 106, 110, 121
Blepharophimosis–Ptosis–Epicanthus inversus, 42

Buffers
  Hepes, 79
  sodium bicarbonate, 79
Burkitt's lymphoma – *see* Lymphoma

Café au lait spots, 109
Campomelic dysplasia, 51
Candida spp., 79
Carcinogens, 115
Carcinoma
  bladder, 32
  breast, 33
  follicular renal cell, 43
  large bowel, 32
  lung, 32
  ovarian, 33
  renal, 32
  squamous cell, 32
  uterine, 33

  thymidine, 78, 80, 122
Centering telescope, 57, 58
Centric segment, 14
Centromere
  DA-DAPI + ve, 2–4
Centromeric
  C-band, 2–4
  division, premature, 81
Cerebellar ataxia, 109
Chang medium – *see* Media
Charcot–Marie–Tooth syndrome, 49
Chemicals
  cytotoxic, 115, 117, 121–122
  exposure limits, 120
  radiomimetic, 81
  risk phrases, 115–116
  testing *in vitro*, 105–106
Chiasmata, 2–4

Chorionic villus sampling (CVS), 90, 91, 94, 97
  direct method/preparation, 90, 91, 97, 98
  incidence of constitutional abnormalities, 5–6
  interpretation of results, 98
  long-term culture, 90, 97
  segregation patterns of Robertsonian translocations, 13
  semi-direct method, 90
  short-term culture, 90
Choroid plexus cysts, 97
Chromatid-type aberrations/damage
  induced, 81, 103, 106, 107, 108
  spontaneous, 106
Chromatin identification
  X, 71
  Y, 72

microdeletion, 42–51, 82
polymorphisms, 2–4
  in banding, 68, 69
puffing, 81
ring, 11, 103
satellites, 3
segregation – *see* Translocation
stalks, 3
submetacentric, 2–4
Chromosome abnormalities
  associated with fetal structural anomalies on scan, 93
  incidence at amniocentesis, 7–9
  incidence at birth, 7–9
  incidence in CVS, 5–6
  incidence in spontaneous losses, 5–6
  incidence of structural, 11
  parental origin, 10
  rates per 1000 at livebirth, 92–93

Chronic granulomatous disease, 52
Chronic lymphocytic leukemia (CLL), 28, 35, 77, 86, 87, 88
  symptoms of, 35
  translocations and oncogenes, 30
Chronic lymphoproliferative disorders chromosome changes in, 28
Chronic myeloid leukemia (CML), 77, 86, 88
Chronic myelomonocytic leukemia (CMML), 23, 87
Clastogens, 105, 115, 116, 121, 122
Clonal rearrangements, 108
Cockaynes syndrome, 107, 108, 109, 110
Colcemid – *see* Mitotic block
Colchicine – *see* Mitotic block
Collagen, 83
Comparative genomic hybridization, 41

Chromomere, 67
Chromomycin A3, 121
Chromosome
  acrocentric, 3–4, 69
  breaks, 103, 115
  centromere (C-bands), 68
  condensation, 116
  deletion, 16
  derivative, 11
  dicentric, 16, 103, 105
  exchanges, 103
  groups, 2–4
  in G-11 bands, 69
  in kinetochore staining, 71
  instability syndromes – *see* Chromosome instability syndromes
  lagging, 102, 108
  marker, 6, 11, 15, 69, 70, 98
  metacentric, 2–4

Chromosome instability syndromes, 107–110
  clinical features, 109
  conditions for expression, 81–82
  methods of analysis, 81–82
  patterns of inheritance, 109
  see also individual syndromes
Chromosome-type aberrations/damage, 102, 103, 104, 105
  induced, 103, 104, 114
Chrondrosarcoma
  extraskeletal myxoid, 32
Chronic granulocytic leukemia (CGL), 86, 87, 88
  chromosome changes in transformed CGL, 25
  classification, 23
  in molecular genetics, 46
  translocations and oncogenes, 31

Complementation studies, 107
Computerized imaging systems, 63
Condensation, inhibition, 115, 116
Condenser
  adjustment, 56
Confined placental mosaicism, 90
Constitutive heterochomatin – *see* Heterochromatin
Containment
  levels, 113
  of cultured samples, 113
  of pathological hazards, 113
Contiguous deletion syndrome, 51
Corpus callosum, agenesis, 97
Counterstains, 115, 121, 122
  distamycin A, 121
  in fluorescence, 64
  methyl green, 122
Coverglass thickness, 60

Craniosynostosis syndrome, 44
Cri-du-chat syndrome, 43
Crossover, in inversion loop, 16
Crown–rump length, 91, 92, 99
Cutaneous T-cell lymphoma – *see* Lymphoma
Cystic hygromata, 97
Cytochalasin B, 103, 121
Cytogenetic endpoints, 102–103
Cytokinesis inhibitor, 121
Cytomegalovirus, 119
Cytotoxicity testing, 106
Cytotrophoblast, 90

Dandy–Walker syndrome, 42
Decontaminants, 113–114
    see also Disinfectants
Dermatofibrosarcoma protuberans, 31
Desmosome complexes, 83

surfactants, 119
Virkon, 114
see also Ethanol
Dispermy, 10
Disseminated intravascular coagulation (DIC), 87
Distamycin A – *see* Counterstains
Distamycin A–DAPI banding – *see* Banding
DNA, 114, 115
    cross-linking agents, 110
    replication, 103, 106, 115
    synthesis, 108, 116
Doliocephaly, 97
Down syndrome, 51, 89, 92, 93, 94, 96
Doxorubicin – *see* Adriamycin
DPX, 120
Duchenne muscular dystrophy, 52
Dulbecco's modified MEM – *see*

Erythrocyte sedimentation rate (ESR), 38
    raised, 88
Erythrocytes, 84, 97
    normal values, 84
Essential thrombocythemia (ET), 23, 86, 87
    symptoms of, 39
Estriol, unconjugated, 89, 95, 96
Ethanediol, 120
Ethanol, 114, 120
Ethidium bromide, 121
Ethyl methane sulfonate, 81
Ethylating agents, 81
Ewing's sarcoma – *see* Sarcoma
Excitation filter – *see* Filter
Extraskeletal myxoid chondrosarcoma, 32
Eyepiece (microscope)

Developmental delay, 109
Dichroic mirror, 62, 64
  in computerized image systems, 64
Diepoxybutane – *see* Alkylating agents
Differential replication staining – *see* Banding
Diffuse large cell lymphoma – *see* Lymphoma
DiGeorge syndrome, 31, 46, 82
Dimethylsulfoxide (DMSO), 79, 80
Disinfectants, 113–114
  activities, 119
  alcohols, 119
  aldehydes, 119
  formaldehyde, 114
  formalin, 114
  glutaraldehyde, 114, 120
  hypochlorite, 114, 119
  phenolic, 119

Media
Duodenal atresia, 97
Dwarfism, 109
Dyspnea, 35, 37

Edema, 97
EDTA, 79, 80
Edwards' syndrome, 92, 93
Effusion, pleural, 97
Electronic darkroom
  in photography, 65
Emission spectra
  of fluorochromes, 61, 73
Eosinophilia, 87
  potential diagnosis, 87
Eosinophilic leukemia, 87
Eosinophils, 84, 87
Epstein–Barr virus, 37, 119
Erythroblasts, 87

adjustment, 55

FAB classification – *see* individual leukemias
Facial
  abnormalities, 109
  clefting, 97
Familial adenomatous polyposis (FAP), 43
Familial renal cell carcinoma – *see* Carcinoma
Fanconi's anemia, 46, 81, 97, 107, 109, 110, 116
Femur length, 90, 100–101
Fetal
  blood sampling, 91
  measurement, 91–92
  measurements correlated with maternal age, 100–101

Fibroblasts, 78, 83
 culture, 82, 108
Fibronectin, 83
Fibrosarcoma
 infantile, 32
Filters
 barrier, 62
  in error correction, 64
 excitation, 62
Fingers
 abnormalities, 97
Fixation, 122
Fixative, 121
Fluorescein isothiocyanate (FITC), 121
Fluorescence *in situ* hybridization (FISH), 40, 41, 103, 105, 121
 application, 42–53
 imaging systems for, 54, 63, 64

distamycin A type, 2–4
folic acid type, 2–4
induction, 78, 79, 115, 116, 121, 122
Fragile X syndrome, 40, 53, 82
Fuller's earth, 116, 117
Fumigation, 114
Fungi
 disinfectants active against, 119
Fungicides, 79
 Nystatin, 79
FX-1 – *see* Media

G-banding – *see* Banding
G-11 banding – *see* Banding
Gene
 T-cell receptor, 106
Genotoxic exposure effects, 103, 105
Germ cell tumor – *see* Tumor

T-cell, 87
symptoms of, 35
Ham's nutrient mixture F10 – *see* Media
Haploid autosomal length, 4
Haploinsufficiency syndrome, 41
Hazards
 pathological, containment of, 112–113, 119
Head
 abnormal shape, 97
 circumference, 92, 100–101
Hematological malignancies
 classification, 17
Hematological values, normal, 84
Hemizygosity, 44
Hemoglobin, normal values, 84
Heparin – *see* Anticoagulants
Hepatitis B

Fluorescence microscope, 61
   in photography, 64
   with Q-banding, 66
   with R-banding, 67
Fluorite lens – *see* Lens
Fluorochromes, 116
   in cytogenetics, 73
Folate
   antagonists, 82
   -free medium, 78
   levels, 78, 82
Follicular lymphoma – *see* Lymphoma
Formaldehyde – *see* Disinfectants
Formalin – *see* Disinfectants
Formamide, 116, 120, 121
Fragile sites
   aphidicolin-type, 2–4
   5′-azacytidine-type, 2–4
   BrdU-type, 2–4

Gestational age, 91
   assessment from CRL, 99
Giemsa, 68, 121
Glioma
   malignant, 33
L-Glutamine, 79
Glutaraldehyde – *see* Disinfectants
Glycerol, 79
Glycoprotein, 83
Goltz-like syndrome, 46
Granulocytes, 87
Greig cephalopolysyndactyly syndrome, 44
Growth deficiency, 109

Hair
   brittle, 109
Hairy cell leukemia (HCL), 28
   B-cell, 86
   categories of high risk sample, 113
   containment, 119
   +delta, 119
   inoculation, 111
   non-A, non-B including C, 119
Hepatomegaly, 34, 35
Hermaphrodites, 53
Herpes virus, 119
Herpes zoster
   in CLL, 35
Heterochromatic region
   centromere, premature separation, 106, 107
Heterochromatin
   constitutive (C bands), 68
   constitutive, uncondensed, 81, 107
   in DA-DAPI staining, 70
Heterozygosity, 45
Hirschsprung's disease, 46

Histiocytes, 37
Histiocytoma
  malignant fibrous, 32
Histiocytosis
  malignant, 27
HIV
  high risk groups, 113
  containment, 119
Hodgkin's disease, 86, 87, 88
  symptoms of, 36
Hoechst 33258, 121
Holoprocencephaly, 45
Human chorionic gonadotrophin
    (hCG), 83, 89, 95, 96
Human transferrin, 78
Humerus length, 91
Hydrops
  nonimmune, 97
Hyperdiploidy, 25

Inversion, 11, 103, 106
  loop, 16
  meiotic behavior, 16
Irradiation – see Radiation
Iscove's medium – see Media

Kallman syndrome, 52
Karyotype
  chorionic villus, 98
  fetal, methods of, 90–91
  human, 2–4
Keratin, 83
Kinetochore staining – see Banding
Köhler illumination, 56

Langer–Giedion syndrome, 45
Large bowel carcinoma – see
    Carcinoma
Large-cell granular lymphocytic

Liebovitz L15 medium – see Media
Ligands, DNA, 114, 115
Linkage analysis, 44, 46, 47, 48, 50, 52
Lipid viruses – see Virus
Lipoma, 33
  atypical, 31
Liposarcoma
  myxoid, 31
Lissencephaly, 49
Listeria monocytogenes
  containment, 119
Lung carcinoma – see Carcinoma
Lymphadenopathy, 34, 36, 37, 38
Lymphoblastoid cell line, 82
Lymphoblasts, 87
Lymphocyte, 84, 87, 103, 107, 112, 116
  culture, 78, 81
  T-, 106
Lymphocytosis, 38, 87

Hyperviscosity syndrome, 38
Hypodiploidy, 25
Hypomelia, 109

ICF syndrome, 81, 82, 107, 109, 110
Ichthyosis, 109
Immunoblastic lymphoma – *see* Lymphoma
Immunodeficiency, 109
In situ hybridization, 116
  fluorescence (FISH) – *see* Fluorescence, *in situ* hybridization
Infantile fibrosarcoma, 32
Insertion, 11
Intercalators, 114, 115
Interphase nuclei
  in chromatin identification, 71, 72
Interphase screening
  in molecular genetics, 42, 43, 45–50

leukemia/lymphoma (LGLL), 28, 86
Leioma
  of uterus, 31
Leishmann, 122
Lens
  achromatic, 59
  apochromatic, 58
  fluorite, 58
  semi-apochromatic, 58
  'Telan', 60
Leukemia
  potential diagnoses, 87, 88
  setting up samples for, 86
  see also individual leukemias
Leukocytes, 84
  total counts, 84
Leukocytosis, 34, 35, 36, 88
Leukoerythroblastic reaction, 36, 87

Lymphoid malignancies, 109
Lymphoma, 86, 87, 88
  Burkitt's, 37
    symptoms of, 37
  chromosome abnormalities in, 27
  cutaneous T-cell, 28, 37, 87
  diffuse large cell, 27
  follicular (in molecular genetics), 47
  follicular large cell, 27
  follicular small cleaved cell, 27
  immunoblastic, 27
  non-Hodgkin's, 27, 31, 36, 86
  T-cell, 27
  T-cell diffuse mixed large and small cell, 27
  setting up samples, 86, 87
  see also Hodgkin's disease (HD)
Lymphopenia, 87
Lymphoproliferative disorders

setting up samples, 86

Macrocytic anemia – *see* Anemia
Macrocytosis, 35
Malaria
  containment, 119
Malignancy, 107
  acquired chromosome abnormalities
    in, 17
  cytogenetic changes in, 18
Malignant
  fibrous histiocytoma, 32
  glioma, 33
  histiocytosis, 22
Marker chromosomes – *see*
  Chromosome
McCoy's 5A – *see* Media
Mean cell volume, raised, 84, 87
Media, cell and tissue culture, 78

Meiosis, 10
  in inversions – *see* Inversion
Melanoma, 87
Meningioma, 33
Mental retardation, 109
Mercury vapor lamp, 63
Mesenchyme – *see* Chorionic villus
  sampling
Methanol, 120, 122
Methyl green – *see* Counterstains
Microcephaly, 109
Microdeletion – *see* Chromosome
Micronuclei, 102–103
Microscope
  adjustment, 55
    for bright field, 55
    for phase contrast, 57
  fluorescent, 61
  in analysis, 54

vinblastine, 122
vincristine, 122
Moebius syndrome, 47
Monocytes, 84, 87
Monocytosis, 87
Monolayer
  cell dispersal of, 79
Monosomy, 12
  double, 14
  interchange, 14–15
Mosaicism
  at amniocentesis and birth, 7–9
  CVS level III, 97
  definitions in CVS and AF, 97
  in CVS and AF, 90, 91, 97
  in Fragile X, 82
  in Pallister–Killian syndrome, 82
  in spontaneous losses and CVS, 5–6
  in Werner's syndrome, 108

Chang, 78
Dulbecco's modified MEM, 78
for culture of
   amniotic fluid, 78
   bone marrow, 78
   chorionic villi, 78
   fibroblasts, 78
   lymphocytes, 78
FX-1, 78
Ham's nutrient mixture F10, 78
Iscove's, 78
Liebovitz L-15, 78
McCoy's 5A, 78
minimal essential medium (MEM), 78
RPMI 1640, 78
TC199, 78
Medium 199 – *see* Media
Megaloblastic anemia – *see* Anemia

Miller–Dieker syndrome, 49
Minimal residual disease, 42, 43, 45, 46, 47, 48, 50
Mitogens, 79, 122
   B-cell, 79, 80, 82
   PHA, 79, 86, 122
   PWM, 79, 86, 122
   T-cell, 79
   12-*O*-tetradecanoyl phorbol-13-acetate (TPA, PMA), 80, 86, 115, 122
Mitomycin C – *see* Alkylating agents
Mitotic
   arrest, 79
   block, 116, 121, 122
      colcemid, 79, 86, 121
      colchicine, 121
      podophyllotoxin, 122
   spindle inhibitors, 115

percentage exclusion, 98
Multinucleates, 83
Multiple myeloma, 86, 87, 88
   chromosome changes in, 27
   setting up samples for, 86
   symptoms of, 37
   translocations and oncogenes, 30
Mustine hydrochloride – *see* Alkylating agents
Mutagens, 115
Mutation analysis, 44, 46, 47, 49, 50, 52
Mycosis fungoides
   chromosome changes in, 28
   setting up samples for, 87
   symptoms of, 37
Myeloblasts, 87
Myelodysplastic syndromes (MDS), 86, 87, 88

chromosome changes in, 22, 23
FAB classification of, 22
potential diagnoses, 87–88
setting up samples for, 86
symptoms of, 35
see also individual myelodysplastic
syndromes
Myelofibrosis (MF), 23, 86, 88
Myeloma – see Multiple myeloma
Myelomonocytic leukemia, 87
Myeloproliferative disorders (MPD),
86, 87, 88
chromosome changes in, 24
classification of, 23
potential diagnoses of, 87–88
setting up samples for, 86
see also individual
myeloproliferative disorders

Normochromic/normocytic anemia –
see Anemia
Nuchal thickening, 90, 91, 97
Nucleolar organizer regions
in silver staining, 69
Nucleoside analogs, 114, 115, 116

Objective (microscope)
coverglass thickness with, 60, 61
lens
achromatic, 59
apochromatic, 58
fluorite, 58
lens, body of, 57
lens, markings of
color correction, 59
flatness of field, 59
low power, 55
magnification, 58

Phytohaemagglutinin (PHA) – see
Mitogens
Piebald trait, 43
Placental biopsy, 90–91
Plasma cell leukemia (PCL), 28, 86
Plasma cells, 37, 38
Plasma volume, 84
Platelets, 23, 35, 38, 39, 84, 87, 88
normal values, 84
Pokeweed mitogen – see Mitogens
Polycythemia rubra vera (PRV), 23,
86, 87
symptoms of, 38
Polydactyly, 97
Polymerase chain reaction (PCR)
in molecular genetics, 40, 42–49, 50,
52, 53
Polymorphism – see Chromosome
Polyploidy, 115

Myxoid liposarcoma, 31

Near haploid, 25
Neoplasia, 87, 109
Neuroblastoma, 33
Neurofibromatosis, 50
Neurological abnormalities, 109
Neutropenia, 87
Neutrophil, 84, 87
  leukocytosis, 87
Neutrophil alkaline phosphatase (NAP), 35
Nijmegen breakage syndrome, 81, 107, 109, 110
Nitrogen mustard – *see* Alkylating agents
Non-Hodgkin's lymphoma (NHL) – *see* Lymphoma
Nonlipid virus – *see* Virus

mechanical tube, length of, 59, 60
  phase contrast, 57
  resolution, 57
Olivomycin, 122
Omphalocele, 97
Ovarian carcinoma – *see* Carcinoma

Packed cell volume (PCV), 84
Pallister–Killian syndrome, 82
Pancytopenia, 35, 87, 109
Paracentromeric regions, 81
Patau syndrome, 92, 93
Peripheral neuroepithelioma, 32
Phase contrast – *see* Microscope
Philadelphia chromosome, 17, 27, 29
Phorbol esters – *see* Mitogens
Photography, 64
  equipment and materials, 74
Photosensitivity, 108, 109

Population monitoring, 103, 105
Postzygotic error, 10
Prader–Willi syndrome, 40, 48
Prenatal diagnosis, 89–101
  screening methods, 89–91
    maternal age, 89
    maternal serum biochemistry, 89
    ultrasound, 90
Prolymphocytic leukemia (PLL), 28
  B-cell, 86
  T-cell, 87
Propidium iodide, 122
Prostatic adenocarcinoma, 33
Pseudomosaic/ism, 9, 90

Q-banding – *see* Banding
Quadriradials, 106
Quadrivalent
  segregation in reciprocal

translocations, 114–115
Quinacrine
dihydrochloride, 122
mustard, 122

Radiation
dose assessment, 103–104
UV, 82, 108, 110
X-ray, 81, 106, 110
Radius, aplasia of, 97
R-banding – *see* Banding
Reagents, 79
Red blood cells – *see* Erythrocytes
Red cell mass, 84
Refractory anemia (RA), 22, 87
Refractory anemia with excess blasts
(RAEB), 22, 87
Refractory anemia with excess blasts
in transformation (RAEBT), 22,

114–115
Roberts syndrome, 81, 107–108, 109,
110
Rouleaux formation, 37
RPMI 1640 – *see* Media
Rubella
containment, 119
Rubeola
containment, 119
Rubinstein–Taybi syndrome, 49

Saethre–Chotzen syndrome, 44
Salivary gland adenoma, 32
Salt solutions, balanced (BSS), 78
Eagles' BSS, 78
Hanks' BSS, 78
Sarcoma, 109
Ewing's, 32
synovial, 31

Sodium
bicarbonate – *see* Buffer
hypochlorite – *see* Disinfectants
Somatic cell hybrids
in G-11 banding, 69
Southern blotting, 43, 44, 46, 47, 51–53
Soybean lipid, 78
Spherocytosis syndrome, 45
Spindle inhibitors – *see* Mitotic spindle
inhibitors
Splenic enlargement, 88
Splenomegaly, 34, 36, 38, 39
Squamous cell carcinoma – *see*
Carcinoma
SRY, 53
Staining, block, 66, 103
Stains, 115, 121–122
see also individual stains
Steroid sulfatase deficiency, 52

Refractory anemia with ringed sideroblasts (RARS), 22, 87
Renal anomalies, 97
Renal carcinoma – *see* Carcinoma
Replication studies, 116, 121
Restriction endonuclease Giemsa staining – *see* Banding
Restriction enzymes
  in banding, 68
Restriction fragment length polymorphisms (RFLP), 42–45, 48–52
Retinoblastoma, 33
  in molecular genetics, 47
Reverse transcription (RT), 42–50, 52
Rhabdomyosarcoma, 31
Ringed sideroblasts, 22, 35
Risk phrases for hazardous chemicals, 87

Satellites – *see* Chromosome
Segregation – *see* Translocation
Semi-apochromatic lens – *see* Lens
Serum, 78
  albumin, bovine, 78
  biochemistry, maternal, 89–90, 95, 96
  immunoglobulins, 88
  paraprotein, 88
Sézary's syndrome, 28, 87
  chromosome changes in, 28
  symptoms of, 37
Short stature, 109
Sideroblasts – *see* Ringed sideroblasts
Silver staining – *see* Banding
Sister chromatid exchange, 81, 102, 103, 106, 108, 110, 114
Skeletal abnormalities, 109
Smith–Magenis syndrome, 50

Streptonigrin, 81
Synchronization – *see* Cell synchronization
Synchronization agents – *see* Cell synchronizing agents
Syndactyly, 97
Synovial sarcoma – *see* Sarcoma
Syphilis
  containment, 119

Tallysomycin, 81
T-banding – *see* Banding
T-cell
  adult T-cell leukemia, 28
  diffuse mixed large and small cell lymphoma, 27
  lymphoma, 27
  receptors, 27
  translocations, 30

Telangiectases of conjunctiva and skin, 109
Telangiectatic erythema, 109
Teratogens, 115, 116
Tetanus
  inoculation, 111
Tetraploid/y, 6, 9, 98
α-Thalassemia, 48
Thymidylate stress, 82
T-lineage
  ALL, 24
  CLL, 28
Thrombocytopenia, 35, 36, 37, 39, 87
Thrombocytosis, 88
Thumb, asplasia, 97
Toxic substances
  classes, 114-115
  safe use, 117–118
  uses in cytogenetics, 115–117

and oncogenes in malignancy, 29–31
Trichorhinophalangeal syndrome, 45
Trichothiodystrophy, 107, 108, 109, 110
Trinucleotide expansion, 53
Triploid/y, 6, 10, 97, 98
Triradial, 104
Trisomy
  13, 10, 13, 95, 97, 98
  see also Patau syndrome
  18, 10, 95, 97, 98
  see also Edwards' syndrome
  21, 10, 12, 13, 94, 97, 98
  see also Down syndrome
  autosomal, 98
  double, 6, 9
  tertiary, 14, 15
Trophoblast, 83
Trypsin, 79

Uterine carcinoma – see Carcinoma

Van der Woude syndrome, 42
Variants – see Chromosome polymorphism
Ventricular septal defect (VSD) – see Cardiovascular disorders
Vermiculite, 116, 117
Versene – see EDTA
Vimentin, 83
Vinblastine – see Mitotic block
Vincristine – see Mitotic block
Virkon – see Disinfectants
Virus, 119
  disinfectants active against, 119
  lipid, 119
  nonlipid, 119
Vitamins, 78

Toxicology testing, genetic, 105
Transformation, 116
Transforming agents, 115
Translocation, 12–15, 103, 105, 106
  reciprocal
    3:1 segregation, 14–15
    4:0 segregation, 14
    adjacent 1 segregation 14–15
    adjacent 2 segregation, 14–15
    alternate segregation, 14–15
    quadrivalent segregation, 14–15
  Robertsonian, 12–13
    3:0 segregation, 12
    adjacent segregation, 12
    alternate segregation, 12
    segregation patterns at amniocentesis and CVS, 13
    trivalent segregation, 12
Translocations

Tuberculosis
  containment, 119
  inoculation, 111
Tumor
  benign ovarian, 33
  germ cell, 33
  promoters, 115
  see also Askin tumor *and* Wilms tumor
Tumorigenesis, 115
Turner syndrome, 9, 92, 93

Ultrasound – *see* Prenatal diagnosis
Ultra-violet radiation (UV) – *see* Radiation
Unconjugated estriol – *see* Estriol
Uniparental disomy, 48
Uridine, 80
Uropathy, obstructive, 97

WAGR, 47
Waldenström's macroglobulinemia, 86, 88
  chromosome changes in, 28
  symptoms of, 38
Waste disposal, 114
Werner's syndrome, 108, 109, 110
White cell
  count, 84
  increase – *see* Leukocytosis
  primitive, in blood, 83
Williams' syndrome, 44
Wilms tumor, 32, 47
  in molecular genetics, 47
Wolf–Hirschhorn syndrome, 43

X chromatin identification – *see* Banding
Xeroderma pigmentosum, 82, 107,

108, 109, 110
X-rays – *see* Radiation
45, X, 6, 8, 9, 10, 97
 see also Turner syndrome
XXX, 6, 8, 9, 10, 93
XXY, 6, 8, 10, 93

XYY, 6, 8, 9, 10, 93
Xylene, 120

Y chromatin identification – *see* Banding

Yeast artificial chromosome (YAC), 41

Zellweger syndrome, 45
Zfy, 53

Keep up-to-date with looseleaf laboratory manuals from Wiley...

OUT NOW!

PROCEDURES IN ELECTRON MICROSCOPY

Principal Editors:
A.W. Robards,
Institute for Applied Biology,
York, UK, and
A.J. Wilson, Centre for Cell & Tissue Research, York, UK.

ISBN: 0-471-92853-4
Published April 1993   1400pp

Updates: Two pa, approx 150pp each,
£95.00\US$155.00

Price: £350.00\US$575.00
(includes 1994 updates).

PREPARATIVE BIOTRANSFORMATIONS: WHOLE CELL AND ISOLATED ENZYMES IN ORGANIC SYNTHESIS

Editor: Professor S.M. Roberts
Associate Editors: Dr K. Wiggins
and **Dr G. Casy,** University of Exeter, Exeter, UK.

ISBN: 0-471-92986-7   1992   500pp

Updates: Two pa, approx 100pp each,
£70.00\US$105.00

Price: £275.00\US$420.00
(includes 1994 updates).

CELL & TISSUE CULTURE: LABORATORY PROCEDURES

Principal Editors:
A. Doyle, J.B. Griffiths and **D.G. Newell,**
Centre for Applied Microbiology and Research, Porton Down, Salisbury, Wiltshire, UK.

ISBN: 0-471-92852-6
Published July 1993   1200pp

Updates: Three pa, approx 100pp each,
£95.00\US$155.00

Price: £350.00\US$575.00
(includes 1994 updates).

Order from: **Looseleaf Department, John Wiley and Sons Ltd,**
Baffins Lane, Chichester, West Sussex PO19 1UD, UK.
Telephone: (0243) 843251 (UK only), +44 (243) 843251 (Outside UK)
Fax: (0243) 820250 (UK only), +44 (243) 820250 (Outside UK)

ESSENTIAL DATA SERIES

All researchers need rapid access to data on a daily basis. The *Essential Data* series provides this core information in convenient pocket-sized books. For each title, the data have been carefully chosen, checked and organized by an expert in the subject area. *Essential Data* books therefore provide the information that researchers need in the form in which they need it.

Centrifugation
D. Rickwood, T.C. Ford & J. Steensgaard
0 471 94271 5, March 1994, £12.95/$19.95

Gel Electrophoresis
D. Patel
0 471 94306 1, March 1994, £12.95/$19.95

Light Microscopy
C. Rubbi
0 471 94270 7, April 1994, £12.95/$19.95

Vectors
P. Gacesa & D. Ramji
0 471 94841 1, September 1994, £12.95/$19.95

Human Cytogenetics
D. Rooney & B. Czepulkowski (Eds)
0 471 95076 9, October 1994, £12.95/$19.95

Animal Cells: culture and media
D.C. Darling & S.J. Morgan
0 471 94300 2, due October 1994, £12.95/$19.95

Nucleic Acid Hybridization
P. Gilmartin
0 471 95084 X, due 1994, £12.95/$19.95

Enzymes in Molecular Biology
C.J. McDonald (Ed.)
0 471 94842 X, due 1995, £12.95/$19.95

ORDER FORM

Please send me:

Qty	Title	Price/copy	Total
......	....	....	....
......	....	....	....
......	....	....	....

All prices correct at time of going to press but subject to change. Your order will be processed without delay, please allow 21 days for delivery. We will refund your payment without question if you return any unwanted book to us in re-saleable condition within 30 days. All books are available from your bookseller.

Method of payment

☐ Payment £/$_____ enclosed (payable to John Wiley & Sons Ltd).
Orders for one book only – please add £3.00/$4.95 to cover postage and handling. Two or more books postage FREE.

☐ Purchase order enclosed

☐ Please send me an invoice
(£3.00/$4.95 will be added to cover postage and handling)

☐ Please charge my credit card account
    ☐ American Express   ☐ Diners Club
    ☐ Visa   ☐ Mastercard

Card no. |_|_|_|_|_|_|_|_|_|_|_|_|_|_|_|_| Expiry: |_|_|_|

Signature: _____

Telephone our Customer Services Dept with your cash or credit card order on 01243 829121 or dial FREE on 0800 243407 (UK only)

Send my order to:

Name (PLEASE PRINT) _____

Position: _____

Address: _____

_____

_____

_____

Telephone: _____

Signature: _____ Date: _____

Return to: Rebecca Harfield, John Wiley & Sons Ltd, Baffins Lane, Chichester, West Sussex PO19 1UD, UK. Telefax: (01243) 539132
or: Wiley-Liss, 605 Third Avenue, New York, NY 10158-0012, USA. Telefax: (212) 850-8888

☐ If you do not wish to receive mailings from other companies please tick this box or notify the Marketing Services Department at John Wiley & Sons Ltd.

⊛ WILEY